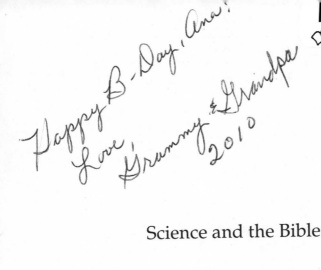

Science and the Bible

Science and the Bible
Volume 3

30 Scientific Demonstrations
Illustrating Scriptural Truths

Donald B. DeYoung

Baker Books

A Division of Baker Book House Co
Grand Rapids, Michigan 49516

© 2002 by Donald B. DeYoung

Published by Baker Books
a division of Baker Publishing Group
P.O. Box 6287, Grand Rapids, MI 49516-6287
www.bakerbooks.com

Fourth printing, May 2008

Printed in the United States of America

Library of Congress Cataloging-in-Publication Data
DeYoung, Donald B.
    Science and the Bible : 30 scientific demonstrations illustrating scriptural truths / Donald DeYoung.
        p.   cm.
    ISBN 10: 0-8010-3023-4 (v.1)
    ISBN 978-0-8010-3023-9 (v.1)
    ISBN 10: 0-8010-5773-6 (v.2)
    ISBN 978-0-8010-5773-1 (v.2)
    ISBN 10: 0-8010-6421-X (v.3)
    ISBN 978-0-8010-6421-0 (v.3)
    1. Bible and science—Miscellanea. 2. Activity programs in Christian education. I. Title.
BS652.D488 1994
220.8′5—dc20
                                                        93-21085

Contents

List of Demonstrations

1. The actual size of the sun is measured with a ruler.
   Genesis 1:16   *The sun displays God's majesty.*
2. The design of a feather is explored.
   Genesis 1:20   *Feathers show God's design.*
3. Acids and bases are distinguished.
   Genesis 1:28   *The universe is orderly.*
4. A mixture of salt and ice results in a very low temperature.
   Genesis 2:1   *God's laws are dependable.*
5. An artificial tree is made from paper.
   Genesis 2:9   *Creation is meant to be enjoyed.*
6. Sound vibrations are made visible.
   Genesis 4:21   *Music is a gift from God.*
7. A map is cut out to show continental separation.
   Genesis 7:11   *The Genesis flood was worldwide.*
8. Sand grains are counted.
   Genesis 22:17   *God's family is large and still growing.*
9. A stairway is built from blocks.
   Genesis 28:12   *We are always in touch with heaven.*
10. Copper is plated onto a nail.
    2 Kings 6:17   *A host of angels defends the believer.*
11. A magnifying lens is made with a water droplet.
    Job 36:24   *Creation evidence is always before our eyes.*
12. Numbers in nature are explored.
    Job 38:4   *Created patterns show God's fingerprint in nature.*
13. The elliptical paths of planets are drawn.
    Psalm 19:1   *Planets obey God's laws of motion.*
14. A lens is shown to invert objects.
    Psalm 94:9   *Eyesight is a precious gift from God.*

Introduction

This book is the third in a series of Bible-science object lessons. Science activities are an excellent way to hold the attention and interest of all ages. The thirty demonstrations in this book include participation by a group, whether two or two hundred. Each lesson is designed to teach a practical Christian truth.

This volume gives special attention to the details of creation. Genesis, the foundational book of Scripture, relates the creation story. These lessons include discussion of evidences and results of creation. The majority of the verses studied are from the Old Testament, although the New Testament is also represented.

Some basic rules for successful object lessons are worth reviewing.

Ten Hints for Successful Science Demonstrations

1. Don't let demonstrations "steal the show." Start with a presentation of the Scripture, memorized if possible. Emphasize the main point of the lesson at the conclusion so the audience will clearly remember it.
2. Practice the science activity ahead of time. Repetition helps bring a smooth delivery, and practice prevents surprises when you are in front of the group. Remember the five Ps: Prior practice prevents poor presentations!
3. Double-check that all needed materials are present and arranged in convenient order. Small details add up to a confident and effective presentation.
4. Adapt demonstrations and Bible lessons to your own situation and talents. Improvise with available materials;

insert new ideas of local or current interest. Creativity will hold the attention of your listeners.

5. When unexpected results occur in a demonstration, laugh and build them into your presentation. The audience will understand and be on your side.

6. Read the background of the Scripture passage. If you are comfortable and familiar with the Bible story, your confidence will be apparent.

7. Good demonstrations use everyday materials. When seen again months later, these items can trigger a memory of the Bible lessons. Use of common items may also encourage the audience to try the demonstrations for themselves, extending the learning process.

8. Many of the best demonstrations involve a dramatic point: an unexpected result that brings "oohs" and "ahs." Science demonstrations should be alive and exciting in this way.

9. Have the audience participate as much as possible. Instead of the lecture approach, help the listener be a part of the Scripture lesson and demonstration.

10. Safety for you and the audience is the highest priority in any science activity. Plan ahead for possible problems; don't take chances. Wear eye protection when appropriate. Know where a first aid kit is located. If the demonstration involves a flame, have water or a fire extinguisher nearby.

I hope you enjoy these Bible-science demonstrations. Further lessons are available in volumes 1 and 2 of this series. It is my hope that these studies will enrich your understanding of the earth and the Creator.

1

How Large Is the Sun?

Theme: The sun displays God's majesty.

Bible Verse: *God made two great lights—the greater light to govern the day and the lesser light to govern the night. He also made the stars* (Gen. 1:16).

Materials Needed:

Ruler with millimeter markings

Calculator

Paper clip or straight pin

Several index cards

Bible Lesson

The creation week was filled with miracle upon miracle as God formed the physical universe. On the fourth day the sun, moon, and stars were set in place. At this time the heavens were filled with a great variety of light sources. The last part of Genesis 1:16 is especially majestic, "He also made the stars." These stars now are known to number in the billions of trillions. In fact, they comprise more than 99.9 percent of the physical creation. The average nighttime star is as big and as bright as the sun. In other words, the sun itself is a star. It dominates our sky and our lives simply because it is much closer to us than the many remote evening stars.

Consider the energy available in our sun. Solar energy is thought to be produced by nuclear fusion reactions. The result is continuous nuclear energy production on a scale vastly greater than all of our power plants combined. In fact, every second the sun gives off much more energy than mankind has produced since the beginning of time. And the sun never stops shining; it is always sunrise somewhere on the earth. We benefit from this solar energy in the form of our pleasant morning light. The sun itself is not running out of energy in any perceptible way; its fuel gauge is still on "full." The significance of Genesis 1:16 is clear: God has infinite amounts of energy available to himself, and he also placed unimaginable amounts in the sun and the stars. God is worthy of all our praise.

Science Demonstration

Participants will measure the actual size of the sun in a simple way. This demonstration must be done during the day and at a time when clouds do not hide the sun. First, a pinhole is punched in the center of an index card. This small hole can be made with a paper clip, pin, or pencil point. On a second index card, draw a small circle with a diameter of 2 millimeters, using a ruler and a pen or thin pencil lead. The circle should be near the center of the card and can be drawn freehand.

Now the punched card is held above the second card as shown in the illustration. Orient the cards in the direction of the sunlight, either while standing outdoors or inside near a window in the path of the sun's rays. A small round image of the sun should appear on the lower card. The pinhole in the upper card acts somewhat like a lens to focus the sunlight. Now adjust the cards' separation until the sun's image is roughly the same size as the 2-millimeter circle drawn on the lower card. While one person holds the cards steady, a second person then measures the cards' separation, also in millimeters. Precision is not needed; let's give this card separation distance the symbol x.

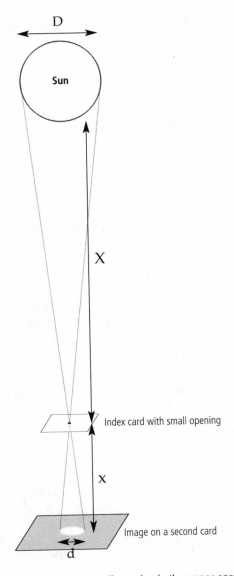

Sunlight passes through a small opening in the upper card, focusing as a small round image on the lower card. Note the narrow triangles with lengths X and x above and below the top card. The drawing is not to scale; the X distance actually is billions of times greater than x.

The actual diameter D of the sun can now be found by substituting the measured value of x into this formula,

$$D = \frac{186{,}000{,}000}{x} \text{ miles}$$

If the x distance is in millimeters, the final answer for the sun's diameter D will be in miles. As a check on the answer, the accurate values are

$$x = 215 \text{ mm (about 8.5 inches)}$$
$$D = 864{,}000 \text{ miles (or 1.4 million kilometers)}$$

Participants will usually get a sun diameter within 10–20 percent of the correct value. Differences are mainly due to the difficulty in accurately producing the 2-millimeter image of the sun on the lower card.

Emphasize the participants' accomplishment: They have measured the vast size of the sun using only a simple ruler! It may be worthwhile to show participants the origin of the formula used. It comes from the similar triangles shown in the figure. The large solar diameter, 864,000 miles, also is typical for many of the nighttime stars. This is a greater distance than many people travel in their entire lifetime. The creation is beyond our understanding, and truly it declares God's great glory and also his great love for us.

Science Explanation

The formula used to calculate the sun's diameter comes from a proportion based on similar triangles. From the figure,

$$\frac{\text{Sun diameter (D)}}{\text{Image diameter (d)}} = \frac{\text{Sun-earth distance (X)}}{\text{Card separation (x)}}$$

Solving,

$$D = \frac{d\,X}{x}$$

The sun-earth separation X averages 93,000,000 miles. If the image size d is adjusted to 2 millimeters, then

$$D = \frac{2mm \ (93,000,000 \ miles)}{x}$$

$$D = \frac{186,000,000 \ miles}{x}$$

where x is measured in millimeters. Notice that the solar image size does not depend on the size of the pinhole but only on the distance between the cards.

One complicating factor is that the sun is about 2 percent closer to the earth during December and January, and 2 percent farther away during June and July. This variation in distance is due to the earth's elliptical orbit. A slightly smaller sun diameter might therefore be measured in summer when the sun is more distant from earth, and a slightly larger diameter in winter when the sun is closer. This earth-sun distance variation has little effect on our weather. It is the earth's tilt that causes our seasons, not the small change in earth-sun distance.

The pinhole measurement technique also can be used in the evening to measure the diameter of a bright full moon. The equation for moon diameter D with a 2-millimeter image is

$$D = \frac{477,000 \ miles}{x}$$

The moon's actual diameter is 2,160 miles, or 3,456 kilometers. For this correct answer, the card separation x is 221 millimeters, or about 8.7 inches. Our moon is about 4 times smaller than the earth and 400 times smaller than the sun.

2

Inside a Feather

Theme: Feathers show God's design.

Bible Verse: *And God said, "Let the water teem with living creatures, and let birds fly above the earth across the expanse of the sky"* (Gen. 1:20).

Materials Needed:

Several feathers (often sold in craft stores)

Water

Dish soap

Bible Lesson

This world is filled with the Creator's handiwork, especially seen in living creatures. Our Bible verse describes the first animal life that appeared on earth, including the sea creatures and the birds. They were supernaturally made in great abundance on the fifth day of the creation. These first animals were not primitive ancestors of today's life. Instead they were made in all of their intricate complexity. In particular, birds show God's careful attention to detail. Flight in nature is a marvel of engineering. The smallest flying insect displays a greater ability to take off, maneuver, and land than even our most advanced military aircraft.

One day in 1948, the Swiss engineer Georges de Mestral was walking his dog in the woods. Arriving back home, he noticed cockleburs caught in the dog's fur coat. These are round seed-pods with a prickly surface that readily cling to clothing or animal fur upon contact. Closer inspection shows tiny hooks on the ends of the burrs. From this finding, de Mestral invented the *Velcro* fastener made of tiny nylon hooks and loops. Velcro has been called one of the great inventions of the past century, but this fastener actually has been around since the creation of cockleburs. Bird feathers show a somewhat similar fastener design in the connections of their side vanes. These are explored in our science demonstration. What additional useful ideas remain in nature, awaiting discovery? Surely there are many. God planned all the parts of our world, complete with endless applications and benefits for us to discover and utilize.

Science Demonstration

This demonstration works best if a feather can be given to each person. Packages of colored feathers are often sold in the craft departments of stores. Feathers found in the outdoors probably should not be used since they may not be clean. Note that feathers have a solid, waterproof surface. They provide excellent lightweight insulation. Hold a feather up before the audience and show how the side vanes can be pulled apart in several places. Listen closely and you may hear the individual parts, called barbules, snap loose. The feather no longer looks smooth but is instead disorganized.

Now comes the interesting part. With your fingers, stroke softly upward on the sides of the feather several times. The barbs should reattach to one another and once again become a smooth surface. The fasteners within the feather consist of many tiny hooks that grasp each other, equivalent to man-made Velcro. This unzipping and quick repair of a feather can be repeated almost endlessly. Birds sometimes separate their feathers in a similar way when cleaning or *preening* themselves. Clearly, the Creator's Velcro is very durable.

Additional activities with feathers are also of interest. A drop of water placed on the feather will form a round bead, showing the feather is entirely waterproof. If a small trace of soap is added to the water drop, however, the surface tension is broken and the water quickly soaks through the feather. It is important for birds to keep their feathers oiled and waterproof.

Hold a feather horizontally and blow gently across the top surface. The feather will then tend to pull upward. This *lift* is how birds and airplanes are able to fly. The air movement decreases the air pressure across the top of the feather, causing the upward thrust.

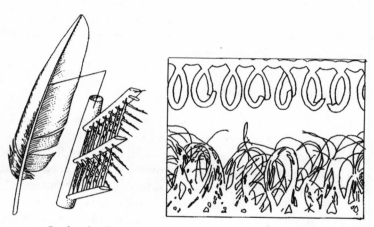

Feather detail   **Velcro detail**

Magnified views of a feather and also Velcro. The enlarged feather detail shows the side barbs that connect together. The Velcro has loops that catch and hold nylon strands when the two surfaces are pressed together.

Science Explanation

A feather's material is made of beta keratin, a fibrous protein. The figure shows a feather and also a piece of Velcro, each magnified about 25 times. The feather's central shaft has side barbs, which in turn have hairlike *barbules*. These are hooked strands that readily grasp each other. Birds also have fluffy *down* feathers that lack the hooks. These small feathers trap air and provide insulation for birds. The cockleburs noticed by Georges de Mestral grow on burdock, a weed that can grow to 6 feet tall. It successfully scatters its seeds by attaching itself to passing animals.

Some scientists believe that feathers somehow slowly evolved from animal scales over a great span of time. In this view, lizard-like animals such as dinosaurs gradually were transformed into birds. However, fossil evidence is lacking. Scales and bird feathers are not at all similar. Each functions uniquely as God intended from the beginning of time.

3

Color Codes

Theme: The universe is orderly.

Bible Verse: *God blessed them and said to them, "Be fruitful and increase in number; fill the earth and subdue it"* (Gen. 1:28).

Materials Needed:

Baking soda

Red vegetable or fruit juice (cooked red cabbage, beets, cherries, cranberries, unsweetened grape juice, etc.)

Vinegar

Bible Lesson

Our selected verse gives God's very first command to mankind. The first part is to produce offspring, and the second is to subdue the earth. *Subdue* is a strong word, and it means to manage, understand, and organize the earth. It includes our responsibility to keep the earth's plants, animals, land, water, and air healthy and productive. After all, this world does not belong to us but to the Lord. Psalm 24:1 says, "The earth is the LORD's, and everything in it." We are therefore stewards, or managers, of God's property.

To care for the earth and utilize its resources successfully, we need to study and understand nature. Many of the great pioneers of modern science were motivated by their confidence that God's world was orderly, predictable, and therefore worthy of inquiry. They saw scientific study as their responsibility, based on Genesis 1:28.

Some Bible critics have complained that our verse leads directly to environmental abuse and ruin. They claim that the word *subdue* implies an irresponsible license to trash, litter, or ruin the earth. But these critics are entirely wrong. Instead of abuse, Genesis 1:28 commands us to help the earth prosper so that it may continue to display God's glory and artistry.

Science Demonstration

The earth's basic materials can be placed into useful categories. These include many liquid chemicals that can be classified as either acids or bases. The acids contain hydrogen and are able to neutralize bases. The bases often feel slippery or soapy; they react with acids to form salts. In the laboratory, various indicators or electronic instruments are used to test liquids to determine their acid or base nature. Many natural indicators are also available. One of the best is the juice from beets or cooked red cabbage. These may not be popular food products, but they are useful in chemistry!

Pour off some of the red juice from a can of beets or cooked red cabbage into a small glass. This juice is a weak acid. Now add about $\frac{1}{4}$ tablespoon of baking soda and stir. The juice color should turn blue or green, which means that the solution now has become basic in nature. The green color indicates a stronger base. If about $\frac{1}{2}$ tablespoon of vinegar is now added, the solution should return to its red color, back to an acid nature. Pickle juice can also be used since it contains vinegar.

Many other natural juices also change color when combined with acids and bases. These include those mentioned under the *materials needed* section, as well as juice from hollyhock plants, rhubarb, and carrot stems. As a science project, you might try testing other colored juices from fruits and vegetables to see how they react with acids and bases. Antacid tablets can also be tested. They are bases that neutralize acids in the stomach.

This activity explores the behavior of acids and bases. The understanding of chemistry is just one part of subduing or understanding the earth and its resources.

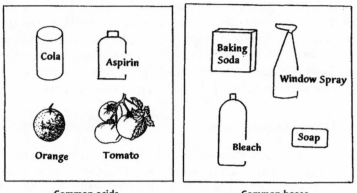

Common acids   Common bases

Common acids include vinegar, citrus juices, sauerkraut, and soft drinks. Bases include baking soda, soap, bleach, cleaning supplies, and milk of magnesia.

Science Explanation

The strength of acids and bases is measured by the level of activity of their hydrogen ions. This is described by their pH number, a chemical symbol that stands for the *power of hydrogen*. Values of pH range from 0 to 15. Acidity is indicated by numbers less than 7. Basic materials, also called *alkalines*, have a pH greater than 8. Neutral materials such as pure water have a pH value of 7 and are neither acidic nor basic. The table lists some pH values for common materials. A change in pH value of 1 means a tenfold increase in acidic or basic nature.

Many vegetable and fruit color pigments function as acid-base indicators. The common *litmus* paper used in chemistry labs is extracted from a lichen plant that grows mainly in the Netherlands. The -*mus* in *litmus* has the same origin as the word moss.

pH

14–	liquid drain cleaner
	oven cleaner
13–	
	limewater (calcium hydroxide, $Ca(OH)_2$)
12–	
	laundry ammonia (ammonium hydroxide, NH_4OH)
11–	
	milk of magnesia (magnesium hydroxide, $Mg(OH)_2$)
10–	detergent
9–	
8–	seawater
	egg whites
7–	pure water (H_2O)
	milk
6–	
5–	coffee
4–	tomato juice
	orange juice (citric acid, $H_3C_6H_5O_7$)
3–	cola (carbonic acid, H_2CO_3)
	vinegar (acetic acid, $HC_2H_3O_2$)
	lemon juice
2–	
	stomach acid
1–	
	bath tile cleaner
0–	

pH values for various common liquids, including several chemical formulas. Materials at both ends of the list are very strong, requiring special care in handling.

4

Counting Calories

Theme: God's laws are dependable.

Bible Verse: *Thus the heavens and the earth were completed in all their vast array* (Gen. 2:1).

Materials Needed:

Quart-sized container

Ice cubes

Salt (water softener crystals or table salt)

Thermometer

Water

Bible Lesson

During the creation week God organized matter and energy in the universe. Light appeared across the heavens, and also a great variety of life was formed upon the earth. Many scientific theories have attempted to explain the origin of the universe. Some of these modern ideas are variously called the steady state, big bang, plasma, and quantum fluctuation theories. However, they all fail to adequately explain the beginning of the universe. We simply cannot understand the creation process because it was supernatural from start to finish. By definition, creation lies beyond our limited, earthly understanding.

The most basic finding in all of nature is called the law of conservation of energy, the first law of thermodynamics. This rule states that energy can be neither created nor destroyed in any process. Energy itself can assume many forms including chemical, heat, nuclear, light, motion, and sound. However, in every process or experiment all of the involved energy can be exactly accounted for. Not a single new calorie of energy ever appears, nor does an existing calorie disappear. This fundamental law of energy conservation makes the universe a dependable place in which to live. Without it we could have no confidence in tomorrow's sunshine or even the adequacy of our next breath. Energy conservation probably was established at the end of the creation week when everything was completed and described as "very good." At this time God ceased inserting physical energy into the universe from his infinite reserves.

Science cannot explain the origin of basic physical laws such as the conservation of energy. People seeking a complete understanding of nature must recognize God's provision of the faithful laws that operate the universe. The Creator clearly has a close connection with each physical detail, including energy.

Science Demonstration

This small-group demonstration shows that energy is constant and always accounted for. It involves the production of a low temperature without refrigeration. You might recognize the procedure as the popular way to make homemade ice cream. For ice cream, of course, additional recipe materials are needed.

A quart container is half filled with ice chips. It will help if large ice cubes are crushed to a smaller size. Snow from outdoors can also be used instead of ice if available. Next add a handful of salt and enough water to make a slush mixture. Stir the mixture, perhaps with a thermometer, and continue to add additional ice and salt as needed.

Keep the thermometer inserted into the brine mixture. You should notice a rapid drop in temperature, well below the normal freezing point of water (32°F, 0°C). If several teams are doing the experiment, they can compete for the lowest temperature attained. Soon the outside of the container will become frosted. A subfreezing temperature as low as 5°F (–15°C) may be reached. If you quickly dip your finger into the slush, you will notice its numbing cold.

The low temperature results from the ice that is forced to melt by the presence of salt. When the ice melts, many calories of energy are absorbed as the water molecules separate into their liquid form. These calories are withdrawn from the surrounding brine solution, thus lowering its temperature. This experiment shows the conservation of energy as calories of energy from the surrounding liquid solution are absorbed in the process of melting the ice.

A mixture of ice, salt, and water results in a very cold temperature.

Science Explanation

The calorie is one of several possible units for measuring heat energy. One calorie will raise 1 gram of water by 1 degree centigrade or Celsius. A diet Calorie, more familiar to us, equals 1,000 of these small water calories. Several other units for heat or energy include metric joules, British thermal units, kilowatt-hours, and foot-pounds.

To melt 1 gram of ice requires about 80 calories of heat energy. This quantity is called the latent *heat of fusion* for ice. In the ice-salt mixture the ice is forced to melt by the presence of the salt. To accomplish this, the ice must withdraw calories from the surrounding brine solution. Theoretically a minimum temperature of -6°F (-21°C) can be reached with an ice and salt mixture if one uses an insulated container. Below this temperature, the ice will coexist with a saturated sodium chloride solution without further melting. This also means that at temperatures below -6°F, it does little good to put salt on winter highways. Calcium chloride, another chemical spread on winter roads, will continue to melt ice down to -40°F (-40°C).

The study of the behavior of materials at very low temperatures is called *cryogenics*. Many new features of creation are revealed as temperatures fall. These include superconductivity, ultrastrong magnetism, and liquefied gases. The lowest possible temperature is called absolute zero, -460°F (-273°C). It results when *all* the calories are removed from a sample, and atomic motion practically ceases. In cryogenics experiments it is possible to get very close to absolute zero, but it cannot be reached completely. Much of outer space has a low temperature close to this cold minimum value.

The term *cryogenics* is also used to describe the freezing of bodies. Some misguided people have themselves frozen at the end of life. Their hope is to be thawed and revived much later when cures are found for aging and death. This certainly is not the Christian approach to eternal life.

5

Pleasant to the Sight

Theme: Creation is meant to be enjoyed.

Bible Verse: *And the LORD God made all kinds of trees grow out of the ground—trees that were pleasing to the eye and good for food* (Gen. 2:9).

Materials Needed:

Old newspapers

Scissors

Rubber bands

Bible Lesson

Trees and other vegetation supernaturally appeared on the earth during the third day of creation. The sun was not made until the next day (see lesson 1), but God nevertheless cared for his newly formed plants and trees. As far as we know, the earth is the only place in the universe where plants grow. There is no convincing evidence of life occurring anywhere else, whether on the moon, Mars, or on any distant planets. Trees were given to us by God to serve many useful purposes. Consider just a few of their benefits:

Food such as apple pie and maple syrup

Building materials and furniture

Many products such as paper and rubber
Shade and shelter for animals and mankind
Medicines derived from trees and plants
Erosion prevention from wind and water

On a more technical note, trees help maintain the health of our atmosphere. All types of vegetation produce oxygen for us to breathe while they absorb carbon dioxide from the air. Vegetation also evaporates or *transpires* moisture into the atmosphere, humidifying the air. A single tree may move hundreds of gallons of water into the air during each growing season.

There is an additional purpose of trees that hasn't yet been mentioned. Our key verse, Genesis 2:9, states that trees were made to be pleasant for us to look at. Whether in summer or winter, there is a beauty and majesty to be seen in trees. God created the earth to be a home for us and also to be a delight before our eyes. We should praise God for his artwork that appears all around us, including the trees.

A camping trip or hike in the woods can be very refreshing. One obvious reason is that we are then surrounded by God's creation. For a short time the daily complications of life are put aside, including its interruptions. Many of us make important life decisions while in such an outdoor setting. In this technological age, we need to occasionally refresh our lives by gazing upon the trees and other details of nature that God makes available to us.

Science Demonstration

Since paper can be made from trees, this demonstration humorously reverses the process and attempts to remake a tree from paper. Begin by rolling 15–20 sheets of newspaper together. Each piece should be half of a newspaper sheet lengthwise, about 14 by 23 inches in size. Newspaper sheets can easily be torn in half at their crease to make this size. Overlap each newspaper piece by several inches as it is placed on the roll. Wrap the sheets somewhat loosely. When completed, the roll can be held together with rubber bands. Now cut $^1/_2$ of the way through the roll from one end in several places with scissors. This is a bit difficult if a greater number of sheets are rolled up; try cutting a few sheets at a time. These cuts will provide the tree branches.

With a large group it is recommended that the roll be almost completed and also cut ahead of time. Then simply add the last couple of sheets and make some final cuts to show the technique during the demonstration. Now announce that you are about to make a tree. Pull slowly upward on the innermost sheets of the roll. The paper should easily pull upward to an impressive height, at least 4–6 feet tall. The scissors cuts will allow the "branches" to spread out and droop downward on all sides. The audience will laugh at the silly appearance of your paper tree, which looks nothing like an outdoor tree, and that is the idea. It is impossible, of course, for anyone to make a real tree. Only God can make a living, beautiful tree.

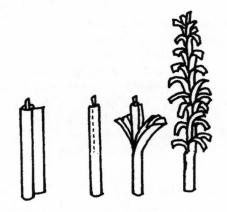

An artificial tree can be made from rolled-up newspaper pages.

Science Explanation

Our word *paper* comes from *papyrus,* a tall plant that ancient Egyptians pressed, dried, and wrote on. Actual papermaking has been traced to Asia, nearly two thousand years ago. In recent centuries, wood pulp replaced rags as the principal source of paper fiber. It was noticed that wasps made paper-like nests by digesting tree and plant material. French scientist Rene Antoine Reaumur described this natural paper production in 1719:

> The American wasps form very fine paper. . . . They teach us that paper can be made from the fibers of plants without the use of rags or linens, and seem to invite us to try whether we cannot make fine and good paper from the use of certain woods.

Paper is just one of the many benefits that result from studying the details of creation.

Over the centuries, the process of papermaking has remained essentially unchanged. First, wood fibers are separated and wetted to make paper *pulp,* or stock. This pulp is thinly spread out and compacted to remove water. Binding materials, fillers, and colors may also be added. Many different grades of paper result from these preparation techniques.

The tallest living things on earth are the Redwood trees of California, some growing to heights of over 360 feet. Trees are also the oldest living organisms on earth. Some of the living bristlecone pine trees of the western United States are more than 4,500 years old. Trees truly are a magnificent part of God's creation and a pleasure to gaze upon.

6

Jumping Particles

Theme: Music is a gift from God.

Bible Verse: *His [Jabal's] brother's name was Jubal; he was the father of all who play the harp and flute* (Gen. 4:21).

Materials Needed:

Pen or pencil

Wide rubber band

Sand or salt grains

Sheet of dark cardboard

Bible Lesson

It is often assumed that people in early times were primitive and uncivilized. Sometimes they are even drawn as apelike or animal-like in appearance. However, the Bible gives a far different picture. From the time of creation, mankind has been given outstanding abilities in thinking and creativity. Tubal-Cain, just six generations from Adam, was a skilled worker of metals (Gen. 4:22). His brother Jubal was a musician who probably designed the instruments that he played. Ever since Jubal, gifted people have had the ability to compose and produce music. The actual instruments mentioned in Genesis 4:21, the harp and flute, are not identified with certainty. The early harp was handheld, and the flute or pipe may have been somewhat like an oboe. Music clearly has been with us since the creation, adding enjoyment to all of our lives. Music is a special gift from God.

Science Demonstration

In general, sound may be defined as a vibration. The vibrating objects may be violin strings, a clarinet reed, or your own vocal chords. Our demonstration makes visible the vibration of small particles. The effect was first seen by scientist Ernst Chladni (1756–1827) two centuries ago.

Lay the stiff, dark cardboard on a table or desk. Hold it in place on the table with a heavy object, and let it extend several inches over the edge. Sprinkle grains of sand, sugar, or salt on the paper surface. The grains can now be made to vibrate in interesting ways. Chladni originally stroked the edge of the paper with a violin bow. We can get the same effect by stretching a wide rubber band along the length of a pen or pencil. Then

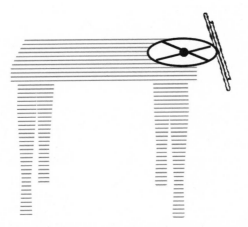

A drawing of Chladni's original experiment. Several of his sand figures are shown also. When a square plate vibrates, sand grains accumulate in the darker areas.

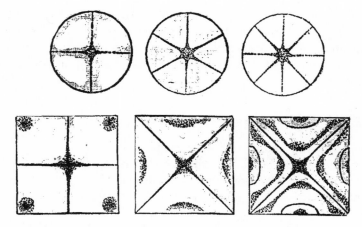

Typical sand patterns obtained on round and square surfaces.

stroke the paper lightly and repeatedly with the rubber band in a downward direction. This takes some practice, but the paper should soon begin to vibrate at various points on its surface. The grains will jump up and down at the locations of maximum vibration. Elsewhere, in quiet areas, the grains will accumulate. Notice how the grains "dance" up and down on the paper surface. The vibration moves from the rubber band to the paper, then to the particles themselves. Invisible air molecules likewise vibrate, carrying sound to your ear.

Science Explanation

Chladni carried out a series of experiments on various shapes of thin metal plates. He observed many complex vibration patterns using sand grains. In this way he was able to explain the different ways that a drum vibrates. Some of his patterns are shown in the figure. Surface areas that vibrate vigorously are called *antinodes*. The quiet regions, lying between the vibrating portions, are called *nodes*. At nodes the grains settle and lie relatively still. It is found that when one portion of a plate moves slightly upward, the adjacent portion moves downward. Thus, adjacent areas vibrate oppositely. Drumheads likewise can be made to vibrate in many different symmetric patterns, depending on how they are struck. Each pattern produces a slightly different sound. It was Chladni who first made these drum vibrations visible. The wooden surfaces of guitars and violins have similar vibration patterns.

7

There Go the Continents

Theme: The Genesis flood was worldwide.

Bible Verse: *In the six hundredth year of Noah's life, on the seventeenth day of the second month—on that day all the springs of the great deep burst forth, and the floodgates of the heavens were opened* (Gen. 7:11).

Materials Needed:

Copies of the map provided

Scissors

Bible Lesson

The world today is very different from the original creation. The continents and the seas themselves were completely reshaped at the time of the great flood of Noah's day. Genesis 7:11 describes the bursting forth of the *springs* or *fountains* of the great deep. This verse describes a worldwide fracturing of the earth's crust. Subterranean water moved upward from ground reservoirs as the rain also fell. The world's climate also changed after the flood from uniformly mild to our current seasonal changes. These physical effects show the global extent of the Genesis flood.

Many scientists reject the idea of a worldwide flood, and they thereby miss a key event in earth history. Many modern theologians also deny the flood story. As a result, they miss the vital message that God is patient with the sin of mankind, but judgment day comes eventually. And when judgment comes, it is complete. In Noah's day, water was used to cleanse the earth. According to 2 Peter 3:12, fire will purify the earth in a future day. This is surely a fearful thought to those who do not acknowledge their Creator. For believers, however, it is an encouragement to realize that someday all things will be made right and faithfulness will be rewarded.

Science Demonstration

Ahead of time, copies should be made of the map figure, perhaps enlarged. It shows the continents and also the locations of many earthquakes. Quakes tend to occur along lines where the earth's crust is fractured, or faulted. Participants are asked to cut out some of the continents and fit them together like puzzle pieces. It is preferable if no instruction is given on which

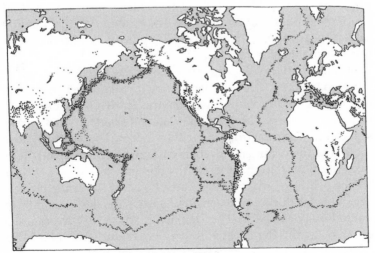

From the National Oceanic and Atmospheric Administration (NOAA).

A map of the world. The dots are the locations of earthquakes, called *epicenters*. They occur along fractures and weak regions of the earth's crust.

continents fit where. Most participants will quickly "discover" the South America–Africa fit. Connections between other land areas are less obvious due to coastal erosion and complex motions of the continents.

Participants may also notice that South America and Africa fit snugly against the fault line in the middle of the Atlantic Ocean. It is helpful if a second copy of the map is available on which to place the cut-out "puzzle pieces." During the flood event the earth's land surface was broken up into the continents we observe today. Continental drift, also called seafloor spreading, thus fits into biblical history. It began dramatically and rapidly with the Genesis flood and still continues today at a much slower pace.

Science Explanation

In 1912, scientist Alfred Wegener first suggested that the earth's continents had split up and drifted apart. His idea was scoffed at and rejected by others for several decades. Today, however, geologists recognize the existence of an original "supercontinent" called *Pangaea*. They believe that about 200 million years ago, Pangaea broke up gradually into the separate continents of today.

In the creation view, the great flood triggered the original Pangaea breakup. The continents were rapidly driven apart due to the worldwide geologic disturbance. The Mid-Atlantic ridge or fault line, shown on the map, remains as one of many great scars from this catastrophic breakup event. The ongoing slow motion of the continents probably is a much-diminished "leftover" motion from the original flood event.

The solid outer part of the earth is called the *crust*. It averages 40 miles thick, and it actually floats on the denser mantle of melted rock beneath. The density of the crust averages 3 gm/cm^3, while the mantle density is greater, around 3.5–5.5 gm/cm^3. The crust is divided into several large plates, similar to puzzle pieces. Some of the edges of these plates are *ridges*, where new crustal material is moving upward from below. Other edges are called *trenches*, where the crust is bending down into the mantle and remelting. The continents are carried along as the plates slowly move, somewhat like conveyor belts. Another illustration of the earth's dynamic crust is the surface of a pot of heated soup. Heat convection causes heated soup to move to the top, then spread outward to the sides of the pot, and again downward. Likewise, new crust material moves upward from the mantle to the earth's surface, then outward.

Satellites measure the exact positions of North America, Europe, and also South America, and find them to be moving apart at the rate of about 1 inch per year. At the time of the great flood, when continental breakup first occurred, the motion would have been greatly accelerated.

8

A Large Number

Theme: God's family is large and still growing.

Bible Verse: *I will surely bless you [Abraham] and make your descendants as numerous as the stars in the sky and as the sand on the seashore* (Gen. 22:17).

Materials Needed:

Small amount of sand

Magnifying glasses (optional)

Small rulers

Bible Lesson

In Genesis 22, Abraham's obedience was tested regarding the sacrifice of his son Isaac. Abraham was willing to obey, and God provided a ram instead of Isaac. Because Abraham did not withhold even his own son, God promised him a great family blessing. Abraham's descendants would number as the stars in the sky and as the sand grains on the seashore. Today, four thousand years later, all believers can be called the "children of Abraham." True to Scripture, no one knows the total number of Christians worldwide and throughout the ages, surely numbering in the billions. God's kingdom is still growing, even in this troubled world, as people acknowledge their Creator and call upon his name.

Science Demonstration

Sand grains provide a good example of an extremely large number. Participants are challenged to count some individual sand grains, and then to estimate the total grains in the entire world. There is some math involved, but the exercise is worth the effort.

Give participants rulers and also a pinch of sand. If sand is not available, salt or sugar can also be used. The particles can be sprinkled on a dark smooth surface that shows them clearly. The goal is to count the number of grains along a 1-inch length of ruler. With the ruler edge, push some of the grains into a line and begin counting. The exact number is not necessary and depends on the particle size. A magnifier may be helpful. My own 1-inch lineup totaled about 100 sand grains.

Now we need to estimate the number of sand grains in a cubic inch, or handful of sand. Take the lineup number and multiply it times itself twice, cubing it. If the initial number was 100, then $(100)^3$ is 1,000,000 or 10^6, 1 million. Going further, a fair estimate for all the sand on all the seashores of the world

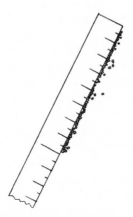

Sand grains are lined up along the edge of a ruler and counted.

is 10^{16} cubic inches, or 10,000 trillion. See the science explanation for the origin of this number. Multiplying these two previous numbers gives the worldwide total of sand grains:

$$10^6 \times 10^{16} = 10^{22}$$

The final number obtained by participants should be close to this result. This number estimates all the sand grains on the earth. Amazingly, it is also approximately the total number of *known* stars in the universe. In words, this number equals 10 billion trillion stars or sand grains. Such a vast number is beyond our comprehension. The meaning of Genesis 22:17 is not that Abraham will have exactly 10^{22} offspring. Instead, the promise is that God's family is beyond counting, and it's still growing!

Science Explanation

Most sand grains consist of quartz fragments made of silicon dioxide, SiO_2. This common mineral has a high hardness and is not easily dissolved or chemically altered. As a result it lies on many beaches in abundance as an end product of rock erosion. Let's estimate the number of cubic inches of sand on all the earth's seashores, 10^{16}, as used in the science demonstration. As a rough guess, assume there are a million miles total of sandy seashores. This number equals 42 trips around the earth and includes the many irregular shorelines. Further, assume the average beach is 100 feet wide and filled 10 feet deep with sand. These would seem to be generous figures. Now convert all three numbers to inches—length, width, and depth—and multiply them to get the total sand volume in cubic inches. There are 63,360 inches in a mile. The final answer is close to 10^{16} cubic inches. As earlier, multiply this number by one million to obtain the total sand grains, 10^{22}.

This is also approximately the total number of photographed stars within all the galaxies of the known universe. This is more than one trillion stars for every person on earth. Whether sand grains or stars—both show God's great glory.

9

A Stairway to Heaven

Theme: We are always in touch with heaven.

Bible Verse: *He [Jacob] had a dream in which he saw a stairway resting on the earth, with its top reaching to heaven, and the angels of God were ascending and descending on it* (Gen. 28:12).

Materials Needed:

    Table or other flat surface

    Rectangular game blocks

        (Dominoes, Jenga blocks, building blocks, etc.)

Bible Lesson

Jacob had recently pretended to be his twin brother, Esau, in order to receive his father, Isaac's, blessing. Esau was very angry about this deception and vowed to kill Jacob. Jacob therefore quickly left home and traveled far to visit his uncle, Laban. Along the way, he rested for the evening in the outdoors, using a stone for a pillow.

During the night, God spoke to Jacob in an unusual dream. Jacob saw a stairway or ladder stretched between heaven and earth. Moving up and down this passageway were angels. God also spoke to Jacob in the dream, comforting him. This vision from God is rich in meaning. The ladder symbolizes Christ, who bridges the gap between heaven and earth (John 1:51). The angels are ministering spirits on earth, active at all times, both day and night. Through prayer and obedience, we thus have access to God in heaven. This is true wherever we are, whether at home or in a wilderness setting like Jacob.

Science Demonstration

This activity works best with individuals or groups of 2–3 people. A model stairway is built by stacking blocks upward, each one offset from those beneath. Lower blocks are gradually spaced outward, with the top block extending outward half its length as shown in the figure. The goal, perhaps as a contest, is to be the first to have the top block *overhang* completely from the bottom block. The result is quite impressive to see, especially if the blocks are stacked at the edge of a table extending outward over open space.

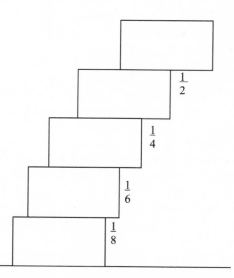

A leaning stairway of blocks, with offsets of $^1/_2$ (top block), $^1/_4$, $^1/_6$, $^1/_8$, etc. No part of the top block is above the bottom block.

Each participant should have 8 or more blocks. Actually, a complete overhang is possible with just 4 blocks above the base block. Have participants continue to stack the blocks, cantilevered outward. The project is fun and also frustrating when an unfinished stack of blocks collapses. This simple block stairway illustrates Jacob's dream. That remarkable stairway extended upward to heaven and assured Jacob of God's presence during his time of need.

Science Explanation

A stack of blocks will balance as long as its *center of gravity* or balance point is supported by the table. The Leaning Tower of Pisa in Milan, Italy, likewise stands because its center of gravity lies above the base, even though the structure is tilted to the side.

A stack of blocks is *just* at the point of tipping when the amount of block overlaps, from the top down, are $1/2$, $1/4$, $1/6$, $1/8$, . . . In this case the center of gravity is positioned directly above the outer edge of the bottom block. This arrangement is shown in the figure. The total overhang is then equal to

$$1/2 + 1/4 + 1/6 + 1/8 + \ldots$$

Mathematicians called this sum a *harmonic* series. Curiously, even though the terms rapidly decrease, their sum increases slowly with no upper limit as more terms (or blocks) are added. In mathematical words, this harmonic series is said to *diverge*. The following table gives some calculated overhang values for stacked blocks, including some large theoretical numbers.

Number of blocks	Possible overhang
1	.5
4	1.04
6	1.225
10	1.46
227	3.1
2.72×10^8	10.1

The table shows that an overhang of 1 whole block is possible with a stack of just 4 blocks above the base. A tower of 227 blocks, probably an impossible feat, could overhang by 3 full blocks. Cantilever bridges are likewise built with beams that project outward toward each other, joining to form a span without need of a center support.

10

An Invisible Army

Theme: A host of angels defends the believer.

Bible Verse: *And Elisha prayed, "O L*ORD*, open his eyes so he may see." Then the L*ORD *opened the servant's eyes, and he looked and saw the hills full of horses and chariots of fire all around Elisha* (2 Kings 6:17).

Materials Needed:

> Lemon juice or vinegar
>
> Nails, paper clips, straight pins, or safety pins
>
> Several pennies
>
> Shallow glass or dish

Bible Lesson

Elisha was a faithful prophet who took upon himself the mantle and testimony of Elijah. Elisha experienced many wonderful instances of God's grace. In 2 Kings 6, a conflict is described between Israel and their opponents, the Syrians or Arameans. The Lord continually showed Elisha the plans of the Syrians ahead of time, thus ruining their strategy of attack. This enraged the king of Aram, and he sent his army to capture the prophet Elisha. Early in the morning, Elisha's servant looked outward from their village and saw the enemy gathered all around them with horses and chariots. He cried to Elisha, "Oh, my lord, what shall we do?" Elisha told his servant, "Don't be afraid." This wonderful phrase of comfort occurs 365 times throughout Scripture. Elisha then prayed that the servant's eyes could see the Lord's protecting army. Immediately the servant saw the hills full of horses and chariots of fire. He then realized the truth of Elisha's words, "Those who are with us are more than those who are with them" (v. 16; see also Ps. 68:17). No battle followed because the enemy was struck with temporary blindness. Instead, Elisha showed mercy to the army that had opposed Israel.

Science Demonstration

Just as the spirit world cannot be seen, individual atoms of matter are also invisible. This activity involves the motion of copper atoms, which can only be seen after many millions of them accumulate. Place 10–20 pennies in a shallow glass or dish. Add enough vinegar or lemon juice to cover the coins, and add a pinch of salt. Since vinegar and lemon juice are weak acids, copper atoms will begin to dissolve from the pennies.

Now other metal items are needed for the copper coating process. Paper clips, pins, or nails work well. These items need to be cleaned first, ideally with scouring powder and water. A clean metal surface is important. The objects are then placed in the penny-acid solution; they can lay directly on the pennies. After 30 minutes, withdraw the metal objects. A slight copper-colored coating should be visible on them. This copper layer is bonded to the metal and should not rub off. If left in the solution

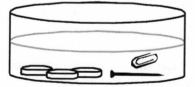

A paper clip, pin, or nail can be copper plated with pennies in a weak acid solution.

for several hours, the metal objects should take on a shiny copper appearance. If bubbles appear on the coins or objects, they can be shaken off. During this *plating* process, millions of copper atoms actually accumulate on the metal objects each second. There is a vast crowd of unseen copper atoms moving between the pennies and the metal. Yet the copper atoms are so plentiful that no loss is noticeable from the pennies. One is reminded of the unseen host of angels that protected Elisha in the Old Testament story. The copper-coated objects can be kept by participants as a reminder of the lesson.

Science Explanation

The number of copper atoms involved in the plating process is truly astounding. Suppose that the completed copper layer on the metal becomes visible at just 0.01 millimeters thick, or 0.004 inches. This represents a thickness of about 1 million atoms. If this thin layer accumulates in 1 hour, then a thickness of nearly 30 new copper atoms must attach to the nail each second. In other words, there is a mighty rush of copper atoms through the solution and to the nail, each atom fitting into a regular crystal structure on the metal surface.

Pennies made before 1982 are mostly copper with some zinc added. From 1982 onward, pennies are made of zinc with only a thin copper coating. The year of the coin does not matter in this exercise, though, since even the newer coated pennies have sufficient copper.

There are several different techniques for the plating process. These include electroplating, anodizing, and chemical plating. Plating with electricity gives more effective results, but the chemical method described here is simpler. Lemon juice is a diluted form of citric acid, $C_6H_8O_7 \cdot H_2O$. Vinegar is acetic acid, CH_3OOH. These acids dissolve the surface copper atoms from the pennies. With a vast number of copper atoms in solution, some find their way to the pin or paper clip surface, where they bond. The electronegativity, or attraction for chemical bonds, is quite similar between copper (1.9) and iron (1.8). Therefore the copper atoms freely move between these metals.

As an alternative exercise, the electroplating process can be shown with a 9-volt battery. Tape paper clips to the terminals as shown in lesson 30. Now touch one wire (positive battery button) to a penny, and the other wire (negative battery button) to the metal object, both submerged in the vinegar or lemon juice solution. You should observe bubbles and relatively rapid copper plating.

11

Through a Lens

Theme: Creation evidence is always before our eyes.

Bible Verse: *Remember to extol his work, which men have praised in song* (Job 36:24).

Materials Needed:

Newspaper page

Needle-nose pliers

Paper clips

Water

Bible Lesson

Job experienced many afflictions from Satan. During this time of testing, three friends or *comforters* came to encourage Job. Another friend Elihu speaks in Job 36. This chapter describes God's control of nature, especially details of the weather. Elihu counsels Job to extol or magnify the works of creation. This word *extol* means to make great, to emphasize, and to concentrate on God's marvelous works. Everyone can see the evidence of God in creation as declared in the next verse, Job 36:25; "All mankind has seen it; men gaze on it from afar." Commentary writer Matthew Henry wrote three centuries ago concerning

creation, "Every man that has but half an eye may see it."
Nevertheless, many people today do not acknowledge God's
presence in the creation. Foolishly they choose to ignore the evidence that surrounds them. Romans 1:20 declares that there is
no valid excuse for ignoring God's works in this way. Life is
filled with choices, and the choices we make have consequences.
We must be sure we choose to honor the Lord who makes all
things, including ourselves.

Science Demonstration

The Dutch scientist Anton van Leeuwenhoek lived from 1632
to 1723. He is known as the father of microbiology because he
applied the microscope to the living world. The instrument had
been invented some decades earlier by Hans Lepershey. Early
magnifiers used lenses made of droplets of water. Participants in
this lesson will make a similar magnifier to illustrate the early
exploration of the microscopic world.

A paper clip needs to be partially straightened. A small
closed loop then is twisted at the end, about $1/8$ inch across,
using needle-nose pliers, as shown. Now dip the loop into water
so that a thin droplet remains suspended within the metal loop.
Hold this loop directly above some small print and observe the
writing through the droplet. The water layer may either increase
or decrease the size of the letters, depending on the droplet's
shape. To make the water drop a magnifier, gently rub the wire
loop across the lip of the water container to remove excess
water.

The water lens magnifies small objects because of the curva-
ture of the water surface. Large permanent lenses of similar
shape usually are made of clear glass or plastic. Scientist van
Leeuwenhoek managed to build simple microscopes by looking carefully through two beads of water at the same time, one
positioned above the other. In this way, with great effort, a
much larger magnification can result. Today, microscopes of
many varieties explore the smallest details of nature. The wonders of creation are found to extend from the vast galaxies to
the smallest atoms.

**Paper clip
magnifier**

Upper concave surface dominates

Diverging or concave lens

Lower convex surface dominates

Converging or convex lens

A paper clip becomes a magnifier when straightened and given a small loop at the end. Shown is a diverging lens that demagnifies and also a converging lens that magnifies an object.

Science Explanation

A lens changes the apparent size of an object by refracting or bending the light reflected from the object. Lenses can be either *converging* or *diverging* in shape.

A diverging or concave lens has at least one surface that curves inward. An object sighted through this lens appears smaller than actual size. In our experiment this decrease in size occurs when extra water is present within the wire loop. The bottom surface will always bulge downward, but the top water surface may be concave. This concave top surface dominates for a diverging lens.

A converging or convex lens has surfaces that are bowed more outward than inward. An object viewed through a convex lens is magnified. In this case the upper water surface in the loop may still be bowed slightly inward, but the bottom downward bulge dominates.

12

Numbers in Nature

Theme: Created patterns show God's fingerprint in nature.

Bible Verse: *Where were you when I laid the earth's foundation? Tell me, if you understand* (Job 38:4).

Materials Needed:

Flowers

Other available items mentioned in the lesson

Pinecones and pine needles

Bible Lesson

Job suffered greatly at the hand of Satan. He lost his children, his health, and his possessions. Most of the Book of Job describes the efforts of friends to comfort him with little success. In chapter 38, God speaks to Job from a whirlwind, asking a series of deep questions. Job is made to realize that God's ways are sometimes "past finding out." In his wisdom, God allows circumstances to happen that we cannot understand from this side of heaven. In our Bible verse, God asks Job if he was present at the creation. Of course, the answer is "no" for Job and for all of mankind. This answer implies that we should not attempt to second-guess exactly how God accomplished creation. In fact, this supernatural event is completely beyond our grasp. The "laying of the earth's foundation" describes God's careful planning of the earth. The next verse describes the "marking off" of the earth's dimensions. That is, mathematical precision was part of the creation details. And still today, arithmetic patterns are found throughout nature. Job did not have all the answers of life, but he knew that God cared for him even more than the physical creation. Job's testimony of trust in God, recorded in Job 1:21, should also be ours. It includes the words "The LORD gave and the LORD has taken away; may the name of the LORD be praised."

Science Demonstration

This activity begins with an arithmetic puzzle. On a sheet of paper, blackboard, or screen, write the following numbers and blanks:

$$1, 1, 2, 3, \underline{\hspace{1cm}}, \underline{\hspace{1cm}}$$

Ask volunteers to guess the next two missing numbers. Some will correctly guess 5 and 8, each found by adding the two previous numbers in the list, $2 + 3 = 5$ and $3 + 5 = 8$. This number sequence continues indefinitely. It is called the *Fibonacci sequence* of numbers, named for the mathematician Leonardo Fibonacci of Pisa who lived eight centuries ago.

Now explain that these particular numbers are very common in nature. They frequently occur in the studies of plants and animals. For example, flower petals often come in clusters of 5 or 8. Flower species with 4, 6, or 7 petals can be found, but they are rare. Pass some objects around the audience or show pictures to a larger group and have them look for Fibonacci numbers. Here are several examples:

Pine needle clusters almost always grow in clusters of 2, 3, or 5 needles.

A typical cloverleaf has 3 petals. Four-leaf clovers exist, but they are mutations or mistakes and are infrequent.

A sand dollar from the sea displays a star pattern with 5
points on its surface. Most starfish also have 5 arms.

An apple cut in half will display a 5-pointed star in its center.

The petals of most flower blossoms, when counted, give the
numbers 5, 8, or higher.

If pinecones are available, the number of distinct spirals
around the outside can be counted. The number is very likely 5,
8, or 13. Why does this Fibonacci number pattern appear in
nature more often than not? There is no evolutionary explanation. Instead, we see a pattern that the Lord chose to imprint
on his works. Creation is not random or accidental but instead
shows intelligent design. Challenge the audience to be on the
lookout for Fibonacci numbers in nature.

Fibonacci numbers are illustrated by 3-leaf clovers, flower
petals, sand dollars, pine needles, and pinecones.

Science Explanation

Leonardo Fibonacci (1175–1230) is also known as Leonardo of Pisa. During medieval times he made many mathematical discoveries. He is best known for the number sequence that bears his name. After the first two numbers, future entries are generated by the formula

$$S_n = S_{n-1} + S_{n-2} \quad n \geq 3$$

That is, the next number is the sum of the two preceding numbers. Fibonacci's initial application of these numbers was to explain the birth pattern of generations of rabbits. This is discussed in many math texts.

The Fibonacci numbers rapidly become large: 1, 1, 2, 3, 5, 8, 13, 21, 34, 55, 89, 144, 233, 377, 610, 987, 1597, and so on. It is more of a challenge to locate the larger numbers in nature, but they do appear. The study of geometric and numerical patterns in plants is called *phyllotaxis*. The following list gives many Fibonacci examples from plants and trees.

Aster	21 petals
Buttercup	5 petals
Chicory	21 petals
Daisy	Spirals in core of blossom number 21 and 34; petals typically number 34, 55, or 89
Delphinium	8 petals
Eastern white pine	Clusters of 5 needles
Enchanter's nightshade	2 petals
Iris	3 petals
Ivy	3 leaves
Larch conifer	Cone has 5 spirals in one direction, 8 in another
Lily	3 petals
Lodgepole pine	Clusters of 2 needles
Marigold	13 petals
Michaelmas daisy	89 petals
Norway spruce	Cone has 3 spirals in one direction, 5 in another

Oxalis	3 petals
Periwinkle	5 petals
Pineapple	Diamond-shaped surface spirals number 8 and 13 in two directions
Plantain	34 petals
Ponderosa pine	Clusters of 3 needles
Primrose	5 petals
Pyrethrum	34 petals
Red pine	Clusters of 2 needles
Sunflower	Spirals of seeds in the flower, depending on the species, number 21 and 34, 34 and 55, 55 and 89, or 89 and 144.
Trillium	3 petals
Virginia creeper	5 leaves
White pine	Clusters of 5 needles

There are also exceptions to Fibonacci numbers in nature, including these:

African violet	4 petals
Clematis	6 petals
Flowering dogwood	4 petals
Honeycomb	6-sided hexagons
Lilac	4 petals
Magnolia blossom	6 petals
Snowflake	6 points

The Fibonacci sequence has many additional interesting properties. For example, the ratio of any two adjacent, larger Fibonacci numbers approaches the *golden mean,* or 1.618. The fraction $^{1597}/_{987}$ is close to this number, which also sometimes is called the *divine proportion.* Rectangular objects with a length-to-width ratio of about 1 to 1.6 are especially pleasing to the eye. The Greek *Pantheon* was built with its length and width based on this ratio. Many breakfast cereal boxes are also designed with these dimensions in mind to attract our attention. It is also no accident that the piano has an 8-note octave with 5 black keys and 8 white keys.

Fibonacci numbers are embedded everywhere in the fabric of art and science. Today there is an international association dedicated to the mathematical study of Fibonacci numbers. Their journal is called the *Fibonacci Quarterly*.

13

Music of the Spheres

Theme: Planets obey God's laws of motion.

Bible Verse: *The heavens declare the glory of God; the skies proclaim the work of his hands* (Ps. 19:1).

Materials Needed:

    Cardboard or corkboard sheets

    Lengths of string

    Sheets of paper, the bigger the better

    Thumbtacks or straight pins

Bible Lesson

Evidence of planning and design in creation surrounds us. From the very smallest objects to the largest, from atoms to galaxies, the fingerprint of God may be clearly seen on all levels. In this lesson we will consider the motions of the planets.

For many centuries the movement of planets across the night sky was a great mystery. These space objects did not follow the regular nightly motion of the background stars. Instead, as weeks or months passed, the planets slowly wandered through the background of stars. The word planet itself means *wanderer*. During the 1600s the German astronomer Johannes Kepler carefully studied the planetary motion of Mars. He discovered several basic rules that govern all planets as they orbit the sun. Today these results are known as Kepler's laws. He was deeply impressed by the regular, predictable motion of the planets. He compared their motion with the regular vibrations of musical instruments. Kepler imagined the planets to be creating celestial music by their regular cycles around the sun. Kepler poetically called planetary motion the "music of the spheres," a term from ancient Greek times. The well-known hymn "This Is My Father's World" refers to this heavenly music:

> All nature sings and round me rings
> The music of the spheres.

At an earlier time it was thought that planetary motion was random and unpredictable. However, Kepler found that the planets move dependably and that they fit the description of Psalm 19:1, just as every other part of the creation does.

Science Demonstration

The curved paths of planets around the sun can be drawn by individuals or small teams of 2–3 people. Place a blank sheet of paper on a cardboard or corkboard surface. Then insert two thumbtacks or pins through the paper and into the cardboard as shown. Tie the ends of a string together to make a loop and place it loosely over the thumbtacks. Now draw a curved line on the paper, constrained by the limits of the string. A smooth egg-shaped curve should result, called an ellipse. An ellipse is somewhat like a flattened circle with two centers at the pin locations that are called focus points.

The planets revolve about the sun in vast orbits with elliptical shapes. The sun is positioned at one of the focus points, shown in the drawing. Participants can experiment by moving the pins closer together or farther apart. These positions respectively give nearly circular orbits and greatly elongated, narrow curves. Ellipse drawing may also be demonstrated on a blackboard. Use small pieces of tape to hold the ends of the string and then draw the curve with chalk.

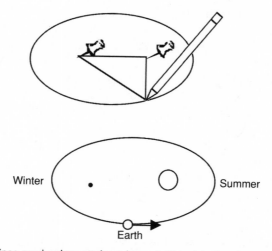

Winter   Summer

Earth

An ellipse can be drawn using a loop of string, two thumbtacks, and a pen or pencil. The lower figure illustrates the earth's orbit around the sun, not drawn to scale. All planet orbits are elliptical in shape.

Science Explanation

All orbiting planets, moons, and comets move along elliptical paths. This is due to the nature of the gravity force. As the illustration shows, planets move closer and then farther from the sun during their annual orbit. The earth is actually a bit closer to the sun during winter in the Northern Hemisphere, 91 million miles, compared to 95 million miles in summer. It is the tilt of the earth that largely controls our seasons, not the distance changes.

Kepler's three laws of planetary motion are expressed as follows:

1. Orbits are ellipses.
2. Planets sweep out equal areas in equal times. That is, they slow down somewhat when farther from the sun (aphelion) and speed up when closer to the sun (perihelion).
3. The orbit time T for a planet varies with its average solar distance R by the relation $T^2 = R^3$.

The ellipses drawn in this activity have various shapes or *eccentricities*. Eccentricity is defined as the separation distance of the focus points divided by the maximum diameter of the resulting ellipse. A perfect circle has zero eccentricity since the two focus points overlap. A narrow ellipse has a high eccentricity, approaching the maximum value of one. The following table lists actual eccentricities for various objects in the solar system:

Solar system member	Eccentricity of orbit
Venus	0.007
Earth	0.017
Moon	0.055
Mars	0.093
Pluto	0.249
Haley's Comet	0.900

The earth's orbit is nearly circular, while that of Pluto is quite narrow and elongated.

14

Seeing Upside Down

Theme: Eyesight is a precious gift from God.

Bible Verse: *Does he who implanted the ear not hear? Does he who formed the eye not see?* (Ps. 94:9).

Materials Needed:

Small magnifying lenses

White cards

Bible Lesson

God has clearly designed the blessing of eyesight. Consider just a few of the details that make our vision possible. *First,* the eye is recessed into its socket and well protected by surrounding bones. Otherwise the sensitive eye surface could be easily injured. *Second,* the eyebrow and eyelash provide an umbrella-like shield from rain and dust. *Third,* our blinking process and eye fluid act like a windshield wiper and washer to keep the outer cornea moist and healthy.

As we enter the interior of the eye, the complexity increases greatly. An image is focused by the cornea and lens upon the retina at the rear of the eyeball. This retina is covered with millions of tiny light sensors called rods and cones. The result is that our eye functions as a color video camera with automatic focusing and clear image storage in the brain. This divinely created mechanism is advanced far beyond any modern video device.

The eye lens itself deserves special attention. Made of clear gelatin-like material, the lens is flexible in its shape. This allows us to focus on nearby objects as the lens becomes thicker and distant objects as the lens slightly thins. This continuous flexing of the lens occurs automatically. As light from an object passes through the lens, it meets the retina in inverted fashion. Thus everything we see is initially upside down. When the visual signal reaches the brain through the optic nerve, it is then corrected so that we can see normally and upright.

Psalm 94:9 reminds us that God formed the eye in his wisdom. Such wonderful senses as sight and hearing do not arise by chance. Further, the One who made our eyes in the first place can surely see all of his creation, including our inner hearts. This should be a fearful thought to those not living for the Lord. For believers, however, it is a precious truth that God knows all our ways and he still loves us beyond measure.

Science Demonstration

We will explore the "upside down" nature of lenses. Each participant needs to be supplied with an inexpensive magnifying lens. The lens is held up to a distant window or light. Then the image is found by using a white card as a small screen behind the lens. Look closely at the focused image on the card and you should notice that it is inverted. If held at some distance, the lens can be looked through directly instead of using the card. Eyes, cameras, and telescopes likewise record an upside down image in this way.

If the lens is held close to an object, for example, to read fine print, the image then will be right side up and enlarged. This magnified image cannot be projected onto a card.

Lenses commonly turn images over because of the laws of optics. Knowing this, God also provided a correcting mechanism within our brains so we can see upright. This vision process works automatically without our concern or understanding. Many of us need the corrective help of glasses, con-

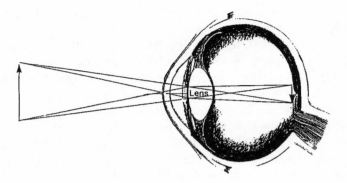

Light passes through the cornea and lens of our eye, then forms an inverted image on the retina.

tact lenses, or eye surgery. Even when eyesight is imperfect, however, it is still a precious gift from God.

Scientific Explanation

The figure shows an eye focused on an arrow. As light rays reflected from the arrow pass through the eye's outer cornea and lens, they are bent or refracted. The result is a clear inverted image on the retina. All the light rays coming from the top of the arrow, for example, meet at the same place on the lower part of the retina.

The human eye lens is flexible, changing shape for different distances. In looking at the moon, for example, the eye lens becomes thinner. In contrast, to view nearby objects clearly the lens becomes slightly thicker. This flexing process is called *accommodation*. Muscles arranged around the lens make this possible. When these muscles become fatigued, it is time to close one's eyes to give them a rest. No cameras are made in a similar fashion with a lens that changes its shape. Instead, cameras rely on lens movement inward or outward, or actual lens replacement. Our eye with its adjustable lens is the most advanced camera known, coming from the hands of the Master Designer.

If time permits, further exploration can be done with two lenses. Hold one close to your eye and the other farther away. Look through *both* lenses at once, with the same eye, at a distant object. This is a challenge but can be done. Move the distant lens outward or inward until a clear image is seen. If the lens close to your eye is thicker than the second lens, the image should be somewhat magnified. What you now have made is one of the great inventions of all time, the refracting telescope. Galileo first constructed such a device from two lenses four centuries ago in 1610. With his primitive telescope, Galileo observed craters on the moon, the rings of Saturn, Jupiter's moons, and many other wonders of the night sky.

15

Rapid Growth

Theme: God cares for his creatures.

Bible Verse: *He makes grass grow for the cattle, and plants for man to cultivate—bringing forth food from the earth* (Ps. 104:14).

Materials Needed:

Sheets of paper of various sizes

Bible Lesson

Psalm 104 has been called the "ecologist's psalm." Ecology is the study of living things and their environments. The psalm wonderfully describes all aspects of creation including springs, trees, storks, wild goats, the moon, lions, sea creatures, and volcanoes. A recurring theme of the psalm is that God cares for the daily needs of his creatures. Verse 14 describes the provision of food both for man and beast.

In 1838, Charles Darwin (1809–1882) read an essay by Thomas Malthus titled "On the Principle of Population." Malthus argued that humans and animals always outgrow their food supply. This then leads inevitably to intense competition and mass starvation. Malthus assumed that populations grow geometrically (2, 4, 8, 16, 32, etc.) while food supplies, at best, grow arithmetically (2, 4, 6, 8, 10, etc.). This idea is falsified today by the dual trends of limited population growth and abundant food supplies. Regional food shortages today are due to conflicts and distribution problems, not a worldwide lack of food. The certainty of future food crises is also denied by Psalm 104:14. Nevertheless, Darwin was strongly influenced by the pessimistic predictions of Malthus. This became the false basis for Darwin's theories of mutation, competition, and natural selection for the formation and improvement of species. In truth, God continues to care for the physical needs of his creatures.

Science Demonstration

The term *geometric* or *exponential* growth is mentioned in the Bible lesson. This type of increase starts out gradually, then rapidly escalates as the numbers continue to double. Charles Darwin misapplied geometric growth, but it is a fascinating type of change. Many items in nature and in our culture increase geometrically:

Computer memory
Food production
Growth of principal with compound interest
Information
Total number of books published

We will demonstrate geometric growth with a paper-folding exercise. Begin by asking participants to estimate how many

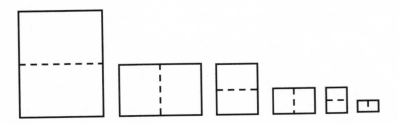

Any size sheet of paper can only be folded 6–8 times. The paper thickness doubles with each fold.

times they can continue to fold in half a sheet of paper. One might suppose the number is almost limitless with an ever smaller result. However, it is practically impossible to fold any paper more than 6–8 times, regardless of its beginning size. This is true of writing paper, newspaper, or tissues. Consider the increasing number of paper thicknesses with repeated folding, 1, 2, 4, 8, 16, 32, 64, 128, 256, 512, 1024, 2048, . . . After just six foldings there are 64 paper thicknesses, about half the thickness of this book, and any further folding becomes very difficult.

Science Exploration

Our geometric growth example can be written as 2^n, where n is the number of paper foldings. Some larger, theoretical numbers of foldings are summarized in the table. Geometric growth can be applied in a positive way to evangelism. Suppose a believer shares the gospel with two other people. Each of these two in turn then share with two others, and so on. The number of people exposed to the gospel will then increase according to the number sequence seen earlier: 2, 4, 8, 16, 32, 64, . . . After just 32 "levels," the entire world population would be reached. In this day of instant e-mail, this possibility of worldwide outreach is exciting.

Number of folds (n)	Total sheets of thickness	Approximate total thickness
5	32	.125 in.
10	1024	4 in.
15	32,768	10.7 ft.
20	1.05 million	341 ft.
25	33.5 million	2 miles
30	1 billion	17 thousand miles
35	32 billion	1 million miles

16

Hot or Cold?

Theme: We cannot always trust in our own judgment.

Bible Verse: *Trust in the LORD with all your heart and lean not on your own understanding* (Prov. 3:5).

Materials Needed:

  Three cups or small containers

  Warm and cold water

Bible Lesson

Think about one of the unwise decisions you have made in the past. We all have such embarrassing moments when we must admit that we were wrong. Our own understanding of any circumstance is imperfect and always subject to error. To trust in the Lord means to not think too highly of oneself. In contrast, God's ways may sometimes seem incomprehensible to us, but he is always trustworthy.

God has blessed modern mankind with much knowledge. In science we have walked on the moon, explored the ocean depths, and split atoms. In the arts, great books and musical pieces have been written. In spite of this progress, however, there is much unhappiness in today's world. Crime, conflict, and tragedy remain with us nearly every day. Clearly, mankind's understanding and control of itself is very limited. But thankfully, God understands us better than we do ourselves. We need to call on him for direction in this life and for the eternal life to come.

Science Demonstration

Each of our five senses—sight, hearing, taste, smell, and touch—is very useful in exploring the world. We depend on them continually for our knowledge and safety. However, these senses can occasionally mislead us. Our lesson demonstration shows how our sense of touch can be confused. It is most easily done while seated around a table.

Three cups are filled with tap water. The first should be quite warm, although not uncomfortably hot to the touch. The second is at room temperature. The third cup contains cold water, perhaps with an ice cube added. Now place your left index finger in the warm water and your right index finger in the cold. Hold them there for about 20 seconds, perhaps while slowly counting to 20. After this, quickly dip both fingers into the room temperature water. The left index finger now will feel like it is submerged in cool water, while the right finger senses that it is in warm water. Yet they are both in the same container. The prior experience of the fingers causes them to falsely sense the normal water temperature. If you didn't know better, you would believe that your fingers were in entirely different containers.

An alternative approach is to blindfold a person initially. Guide them in moving their fingers from the hot and cold containers to the room temperature water. They will be surprised when shown that their fingers are now in the same container.

The obvious lesson is that our senses cannot always be trusted. The sense of touch is valuable, especially when guarding against dangerous temperatures and burns, but it easily can be fooled. Likewise our own understanding of life and events can be wrong. We need direction from God's Word to make the right decisions.

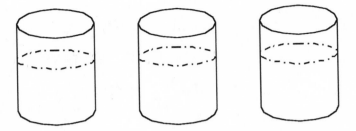

Three containers are filled with warm, normal, and cold water.
They are used to confuse our sense of touch.

Science Explanation

The temperature of an object is a measure of the average motion or kinetic energy of the object's molecules. The more rapidly the molecules move or vibrate, the higher the temperature. The tissue molecules within a finger vibrate thousands of times each second, as if connected with tiny springs. The finger placed in the warm water absorbs some of the heat, which results in a still higher vibration of skin molecules. In the cold water, molecular motion within the finger is slowed. Then when both fingers are shifted to the neutral water, vibrational motion from the warmed finger is given up to the surrounding water. The warmed finger thus loses this energy and feels cool. Meanwhile, the finger from the ice water absorbs vibrational energy from room temperature water and feels warmed as a result. Following the experiment, your sense of touch will soon return to normal.

17

Haste Makes Waste

Theme: It pays to be patient.

Bible Verse: *A man's wisdom gives him patience; it is to his glory to overlook an offense* (Prov. 19:11).

Materials Needed:

Wooden paint stirring stick or thin scrap of wood

Sheets of newspaper

Bible Lesson

Patience can be shown several ways. It includes an understanding attitude, a listening ear, wise council, and forgiveness for others. King Solomon reminds us in Proverbs 19:11 that patience is a sign of wisdom and maturity. Even in Old Testament times, patience was often lacking among people. And on our modern highways especially, violent arguments called "road rage" sometimes arise over trivial incidents and delays. Such people completely lose control of their own emotions and behavior. How sad it is to fill our few days on earth with unnecessary anger and impatience. Such emotions are unhealthy for both the body and the mind.

Let us look to the Lord as our example for patience. Ever since the fall or curse, this world has dishonored God in many ways. Still, God patiently waits for us to be drawn to him. He is willing to forgive our offenses far more than we could ever achieve in forgiving one another. By showing patience toward others, we can reflect God's love to those around us.

Science Demonstration

This demonstration will illustrate the ruin that can result from haste. Place a thin board or stick on a tabletop with nearly half the stick length extending out over the edge. A wooden paint stirrer or similar stick works well. Now place two sheets of newspaper over the stick and smooth them flat. A single newspaper sheet will sometimes tear, so two are recommended.

The idea is to push downward on the protruding stick as a lever, thereby raising the newspaper. Show the group that a slow push easily pivots the stick and lifts much of the newspaper upward as expected. Now ask what will happen if the same task is done quickly, by striking the stick. The usual guess is that the newspaper will rapidly fly upward or perhaps tear itself in half. However, this is not the case. The strongest blow should not budge the newspaper. Instead, the paper will remain flat and the stick will usually break into two pieces. Stand to one

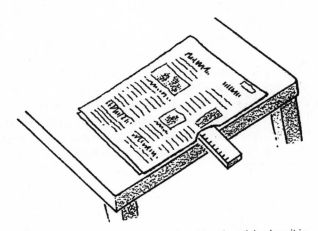

The newspaper refuses to move, breaking the stick when it is struck downward.

side, and take care that the broken stick does not fly through the air toward anyone. Practice ahead of time with extra sticks. Make the concluding point that a slow, patient movement easily lifts the paper. Likewise a patient approach is often the best way to solve problems. Haste makes waste as illustrated by the broken stick!

Science Explanation

This type of activity is sometimes called a *discrepant event*. That is, the results are contrary to one's intuition. It seems almost impossible that lightweight newspaper sheets can withstand the sudden upward force from a wooden stick. Actually, air pressure is involved in breaking the stick. If the covered stick is raised slowly, the paper offers little resistance. Air easily moves beneath the paper, and the pressure above and beneath the paper surfaces remains equal. With rapid motion, however, a temporary thin layer of partial vacuum occurs beneath the newspaper. This results in less air pressure under the newspaper than on its upper surface.

Air pressure at the earth's surface averages 14.7 pounds per square inch (psi). In striking the stick, suppose there results an upper/lower pressure difference of just 1 psi. Also, suppose that the central part of the newspaper comprises about 100 square inches (10 inches by 10 inches). There then results an unbalanced downward force of 100 pounds holding the newspaper down, and the stick therefore quickly breaks. The actual total weight on the entire top of a flat newspaper due to air pressure may be as great as 4,000 pounds, or 2 tons.

Air pressure results from the weight of the earth's atmosphere. Living on the earth beneath this "ocean" of air, we don't often notice its resulting air pressure. This demonstration is one of many dramatic ways to illustrate the air pressure that surrounds us at all times.

18

Water beneath Our Feet

Theme: God supplies our daily needs.

Bible Verse: *All streams flow into the sea, yet the sea is never full. To the place the streams come from, there they return again* (Eccles. 1:7).

Materials Needed:

About 30 marbles

Clear glass or transparent cup

Measuring cup

Water

Bible Lesson

Our lives depend on a continuous supply of fresh water. Think about the last time the water was temporarily turned off in your home. At such times it is surprising how often one automatically reaches for a faucet—with no result. Without water, normal routines quickly come to a halt.

Many people around the world obtain their water from the underground. Rain and melted snow soak into the soil, where they continue to move downward and are slowly filtered by sand and gravel. The resulting water then accumulates within underground layers of rock and gravel, sometimes called aquifers. This water can later be brought to the surface with a pump or open well. Groundwater is available almost everywhere on earth at varying depths. The upper water surface, called the water table, may be only tens of feet deep, or it may be hundreds of feet deep in desert locations.

Our verse describes an apparent mystery involving streams and rivers. How can they keep flowing day after day, even during times of drought? The answer is that water is continually recycled. Water enters streams along their length through underground springs and seepage. Groundwater is constantly on the move, gradually flowing downward by the influence of gravity toward streambeds, lakes, and also to the seas.

Solomon in his wisdom recognized the movement of the earth's water. Evaporation occurs from seas, lakes, and also from the land. Much of this moisture later returns to the earth as precipitation. In this way sea level remains constant, the land is refreshed and watered, and groundwater reserves are recharged. This movement of water, called the water cycle or *hydrologic* cycle, is a precious and essential gift from the Creator.

Science Demonstration

We will explore the amount of water that is stored underground. Teams of 2–3 people can do the activity. A measuring cup is filled to a convenient level with marbles, perhaps 1–2 cups total. It doesn't matter if the marbles are of unequal sizes or if a few extend slightly above the selected measuring line. Now, transfer the measured quantity of marbles to the clear container and fill the measuring cup with water. Pour this water onto the marbles until the liquid level reaches the top of the marbles. Water now fills the openings between the marbles, just as it fills small spaces in the underground. The marbles represent buried sand or gravel. A look at the water remaining in the measuring cup will show how much water has been poured out. Experimenters will find that the spaces between the marbles hold a surprisingly large amount of water. If there was originally an amount of marbles equal to 1 cup, then the water added should be nearly 1/2 cup. This activity helps one visualize how trillions of gallons of water can be stored underground between small pebbles and sand grains.

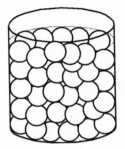

A container filled with marbles, sand, or gravel has considerable additional space available for water.

Science Explanation

The amount of water that can be held by underground materials is called the *porosity*. It is defined as the percentage of the total volume of underground rock or sediment that consists of open pore space. For example, if 0.4 cups of water can be added to 1 cup full of marbles, then the porosity is

$$\frac{.4}{1}(100\%) = 40\%$$

That is, there is 40 percent open space, now filled with water. The following table gives porosity ranges for several common underground materials.

Material	Porosity (%)
soil	35–50
clay	35–80
sand	30–50
gravel	25–40
limestone	0–20
sandstone	0–30
volcanic rock	0–50

It can be seen that most materials hold considerable amounts of water. To extend this activity, the marbles can be replaced by sand or gravel. The resulting porosity values can then be compared with those in the list.

19

As the World Turns

Theme: God turns his world.

Bible Verse: *He sits enthroned above the circle of the earth, and its people are like grasshoppers. He stretches out the heavens like a canopy, and spreads them out like a tent to live in* (Isa. 40:22).

Materials Needed:

Small weight

Length of string

Bible Lesson

Our verse from Isaiah describes how God looks down upon his world and the vast space beyond. The "circle of the earth" either may refer to its round shape, its spin, or to its orbit around the sun. Every 24 hours, the earth turns once on its axis like a giant merry-go-round. This motion provides our sunlit days and dark nights. The earth's rotation also controls our weather. The sun's heat is more evenly distributed by the turning earth, thus avoiding severe hot and cold spots. High- and low-pressure systems and the wide distribution of rain are also connected to the earth's spinning motion.

We may take the earth's rotation for granted, but God does not. He established the earth and its motions for our benefit.

Such details clearly show God's love. There are other planets, but none are prepared for life as the earth is. Truly, there is no place like home.

Science Demonstration

A pendulum can show the earth's turning motion. This project requires careful observation by the participants. The pendulum is simply a string, the longer the better, attached to a high support or the ceiling. It has a compact weight tied near the floor. The pendulum is set swinging with a large arc. Carefully note the direction of the swing. It helps to align the pendulum motion with the edge of a carpet or newspaper on the floor. The pendulum swing will diminish greatly with time, but this does not matter as long as some motion, no matter how small, can still be seen.

Because of the earth's rotation, the pendulum should very slowly change its direction of swing toward the right. In other words, it should gradually move clockwise as you look down upon it. This direction is for the Northern Hemisphere. South of the equator, the directional change is counterclockwise. It is best to set the pendulum swinging in a quiet room without air currents, then check it again after 10–15 minutes. A slight change from its original path should be noticeable. If so, you are see-

As a pendulum swings, it slowly turns to the right. Movement shown is exaggerated.

ing the earth turn! Actually, the direction of the pendulum itself does not change. Instead the floor, the house, and the earth itself all turn slowly beneath the pendulum. As a result the pendulum *appears* to slightly change direction. Somewhat similarly, a ball appears to follow a curved path if it is rolled across a moving merry-go-round.

Science Explanation

The French physicist Jean Foucault first performed this pendulum experiment in 1851. He constructed the pendulum in his cellar, then watched carefully as it slowly altered its direction of swing. Foucault concluded correctly that he was watching the entire earth rotate beneath his pendulum. He also went on to invent the gyroscope, used today in aircraft and space probes. A nonbeliever most of his life, Foucault finally gave honor to his Creator during an illness that took his life at age 48.

The amount of deflection of the Foucault pendulum depends on its location. At the equator, there is no pendulum deflection. In southern states (latitude 30°), the pendulum moves through 7° per hour. In the upper Midwest (latitude 45°), a pendulum moves through about 11° in an hour. The average pendulum motion for these regions, then, is about 1° every 10 minutes. Meanwhile, at the North and South Poles, the pendulum appears to rotate completely once every 24 hours, or 15° per hour. This deflection takes some effort to see, but *is* noticeable and worth the effort. The longer the string and the heavier the weight, the longer time the pendulum will continue swinging, which helps in observing the effect.

The Foucault pendulum is just one application of the *Coriolis* effect, named for a French mathematician. This is the observation that all moving objects north of the equator tend to slowly drift to the right. This includes air masses, rivers, planes, rockets, and even cars. This general deflection is due to the turning earth.

20

A Superabsorber

Theme: Creation is filled with useful gifts.

Bible Verse: *[God] who fashioned and made the earth, he founded it; he did not create it to be empty, but formed it to be inhabited* (Isa. 45:18).

Materials Needed:

Disposable diaper

Two clear glasses

Water

Bible Lesson

Our Bible verse states that the earth is a unique place in the entire universe. It was here that God focused his special attention for our welfare. He supplied the earth with just the right balance of oxygen, water, and useful elements. We are surrounded with examples of obvious design and planning. It is very common today to "put down" any special significance for the earth. After all, it is just one of nine planets circling the sun. Furthermore, we are located in a distant corner of the Milky Way, just one of billions of galaxies. Isaiah 45:18, however, gives an entirely different perspective. After all, it was to the earth that Christ came and walked among men. In this light, the earth is truly a spiritual center of the universe.

Science Demonstration

When God created the earth, he placed within it many potential benefits for mankind. It is the task of scientists and engineers to discover these items in nature and then to develop useful applications. Examples are seen in every list of discoveries and inventions.

We will experiment with one fascinating product called *water absorber*. It may be available from a chemical supplier or landscaper. Alternatively, water absorber powder is found within the fluffy fabric of larger disposable absorbent diapers. Ahead of time, slit the edge of such a diaper and shake out some of the white absorber grains. Very little is needed, less than a teaspoon.

To demonstrate the water-absorbing property, obtain 2 clear glasses. Fill one with water, and place water absorber powder in the other glass, just enough to cover the bottom. Now show the audience what happens when the water is poured back and forth between the glasses. Very quickly the water should mix with the powder and turn into a thick Jell-O-like form that will not pour. The rapid change is quite dramatic. If distilled water is available, it works better than tap water.

Ask the audience to suggest possible uses for the water absorber. A few of them include:

- An absorbent component for disposable diapers. Such a diaper can hold half a gallon of water.
- A gatherer of moisture for plantings.
- The prevention of erosion on hillsides.
- Cleanup of spills and medical wastes.
- Protection of power cables and fiber optics from water leaks.
- Filtering water out of aviation fuel.
- For toys that expand in water.

There are undoubtedly many additional applications for water absorber. This material is one of the countless blessings available to us, originally planned by the Creator.

Water quickly turns solid with absorbing powder.

Science Explanation

Water absorber comes from a group of *hydrophilic* ("water-loving") chemicals called polyacrylates. One of the most common is sodium polyacrylate, which can hold 800 times its weight in distilled water. This polymer chain has sodium atoms along its length that attract and hold tightly on to water molecules. Instead of dissolving in water, the polymer solidifies into a thick gel. The polyacrylate powder and gel are biodegradable and can be disposed of normally.

21

Weighed in the Balance

Theme: God sees our hearts.

Bible Verse: *You have been weighed on the scales and found wanting* (Dan. 5:27).

Materials Needed:

Weighing scale

Any solid palm-sized object, painted gold or silver

Large clear container filled with water

Length of string

Bible Lesson

Belshazzar followed Nebuchadnezzar as the king of Babylon from 553 to 539 B.C. This new king ignored God and instead worshiped man-made idols (Dan. 5:4). During a royal banquet, God's hand wrote a coded message of judgment on the plaster wall for everyone to see. None of the king's men could explain the four strange words that appeared, and Daniel was sought out to interpret the words. The resulting message that God revealed to Daniel was not good news for Belshazzar. His reign was ended, and he was weighed in the balance and found wanting. Also, the country of Babylon was about to fall to its enemies, the Medes and the Persians. All three predictions quickly came to pass, with Belshazzar's death occurring that very night. Belshazzar's "weighing in the balance" meant that God had evaluated his life. Old Testament balances were normally used in the weighing out of payments for goods. Belshazzar's evil life and leadership clearly did not measure up to God's righteous standard. The lesson discussion should include the fact that all of us fall far short of God's perfection. However, Christ in our life can make up for all our failures as we seek his forgiveness.

Science Demonstration

A weighing scale with an easily seen dial is needed, perhaps a small postal scale. The only requirement is that the item to be weighed must be suspended from a string. The figure shows two possible arrangements.

Choose an object that superficially looks like it might be made of silver or gold. This might be a rock or any other solid painted object such as plaster. Show the audience the weight reading of the object as it hangs on the scale and write down the number, rounding it off if necessary. Now place the suspended object, still hanging freely on the string, in a filled vessel of water. The object should not rest on the container bottom. You will notice that with the object submerged, the scale reads a smaller value than before. Objects weigh less when suspended underwater, as you have noticed for yourself when swimming. A comparison of the two scale readings determines what is called the *specific gravity* of the object. Specific gravity (SG) is calculated by the formula

$$SG = \frac{\text{Air weight}}{\text{Air weight-Water weight}}$$

For example, suppose an object weighs 8 ounces in air and 6 ounces in water. Its specific gravity is then found to be $8 \div 2 = 4$. The value of 4 means that the sample is 4 times as heavy as an equal volume of water. The audience can be shown the following table of specific gravities for various items:

Material	Specific Gravity
Cork	.2
Wood	.3–.6
Plastic	.5–2
Water	1.0
Aluminum	2.7
Most Rocks	3–5
Plaster	3–4
Pyrite, Fool's Gold	5.0
Iron	7.9
Copper	8.9
Silver	10.5
Lead	11.3
Gold	19.3

By measuring specific gravity, one can readily determine whether an unknown object is made of gold, silver, or something else. If a plaster object is painted gold, its specific gravity will only be 3–4 instead of 19. Just as this test analyzes unknown objects, God likewise instantly knows our hearts. Belshazzar's life was "weighed in the balance" and found wanting. How do we compare?

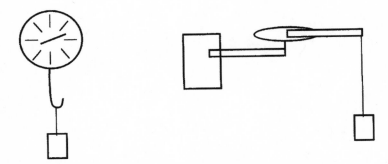

Two ways to weigh a suspended object. (a) Tie the object to a vertical scale and hold it in the air. (b) Tape a light stick support to a pan-type scale, and then tie a string to the stick.

Science Explanation

This measurement technique avoids the need to know the actual volume of the unknown object. Specific gravity is a comparison between the sample weight and an equal volume of water. The quantity called specific gravity has no units because the air and water weights, whether in ounces, pounds, or grams, cancel out. If the units of grams per cubic centimeter (gm/cm^3) are added later to the specific gravity number, the quantity that results is called the *density*. The only difference between specific gravity and density is whether or not the number has units attached. The density of gold is 19.3 gm/cm^3.

Common applications of specific gravity are as an antifreeze test for a car's radiator and also as an acid test for battery fluid. Some of the fluid is drawn into a large syringe. A weight then floats at a particular level in the fluid, depending on the liquid's specific gravity. This determines the antifreeze content of the radiator fluid or the acid concentration of the car battery.

22

A Sure Guide

Theme: The Holy Spirit guides us.

Bible Verse: *But when he, the Spirit of truth, comes, he will guide you into all truth* (John 16:13).

Materials Needed:

Sheets of wax paper, each about 1 foot square

Bible Lesson

Serious hikers often make use of a global positioning system, or GPS. This small handheld computer receives signals from satellites and then calculates a person's exact location. The readout may be either a detailed map or the exact degrees of latitude and longitude for the position. As long as the receiver's batteries hold out, it is calibrated correctly, and the distant satellites are operational, one should not get lost. GPS systems are also useful accessories for boats and cars.

In John 16:1, Jesus said the Holy Spirit would come so that the disciples would not go astray. This promise is more valuable than a GPS system, which is limited to finding geographic locations. Instead, the Holy Spirit helps us to navigate the uncertainties of daily life and also helps us prepare for the future. We are faithfully led through prayer, Bible reading, wise counsel from others, and listening to his voice in our hearts. Just as a

GPS system may guide us across the countryside, the Holy Spirit can guide us safely through the journey of life itself.

Science Demonstration

When you toss a ball through the air, it follows a very predictable path. It moves upward and then back downward along a smooth arc called a *parabola*. To construct a parabola, first give each person a sheet of wax paper. Have them fold the paper horizontally, 2–3 inches above the bottom edge. Then unfold the sheet, and a white line should appear along the crease. Wax paper is used because it produces this permanent fold line.

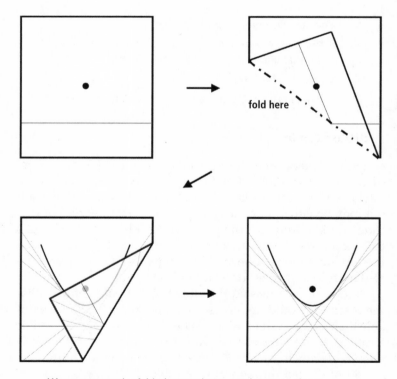

Wax paper can be folded many times, as shown, to give a parabola curve.

Next, a dot is needed, centered on the paper and 2–3 inches above the crease. This dot can be made with one's fingernail or any small blunt object.

Now multiple foldings begin. Starting at one end of the line, fold and crease the wax paper so the original line overlaps the dot. Then repeat this folding at small intervals along the original line. Continue to fold, crease, then unfold the paper, and a smooth parabola will begin to take shape. The final curve is formed by the intersections of all the paper creases. The distance between the initial line and dot controls the actual parabola shape. A small separation gives a steep, narrow parabola. A larger separation gives a broad, shallow parabola.

The resulting curve occurs very generally in nature. The trajectory of an object flying through the air is a parabola. Also, the paths of many comets when near the sun are parabolic in their movement. Just as the Holy Spirit guides us, gravity likewise guides these motions. The shape of a flashlight lens, an auto headlight, and telescope mirrors are also parabolas.

Science Explanation

A parabola is defined as the "locus of points equidistant from a point and perpendicular to a straight line." We have *mapped out* this locus of points by paper folding. With various selections of points and lines, one can fold paper into other mathematical shapes including a circle, an ellipse, or a hyperbola. This exercise is a special category of *origami,* the art of paper folding.

A projectile such as a soccer ball moves through a parabola curve as it rises and falls. This smooth, symmetric shape is controlled by gravity. Some other objects such as golf balls diverge from a true parabola trajectory because of spin and also because of the dimples on the ball's surface. The parabola shape reveals an inherent simplicity and symmetry to most flying objects. They follow the law of gravity established for them. Do we likewise follow the leading of the Holy Spirit as we should?

23

A Clanging Cymbal

Theme: Christians must show love to others.

Bible Verse: *If I speak in the tongues of men and of angels, but have not love, I am only a resounding gong or a clanging cymbal* (1 Cor. 13:1).

Materials Needed:

Cardboard

Four-foot length of string

Masking tape

Paper clips

Scissors

Bible Story

Actions speak louder than words. We have all experienced the truth of this statement. The value of one's testimony may be completely canceled out by a single episode of questionable behavior. In contrast, the display of a godly lifestyle, even with few actual words, can influence many people for good. As Francis of Assisi (1182–1226) said long ago, "Tell others about the Lord; use words if necessary."

Some Bible translations use the older word *charity* in place of love. This is very appropriate since love certainly includes charity, meaning a kind generosity toward others. Lovingkindness toward others expresses the spirit of all the other Christian virtues (John 13:34–35). Thankfully, our key verse shows that love is not determined by our speaking ability. This verse had a special meaning for the New Testament people in the city of Corinth, where speaking eloquence and formal debates were greatly admired and practiced.

Chapter 13 of 1 Corinthians could be called a "hymn dedicated to love." Its description of Christian love is both beautiful and comprehensive. The illustration of a gong or cymbal in our key verse is instructive. Sounds from these instruments are loud and impressive, but they also are very temporary, or transient. As soon as we hear the sound it is gone, except perhaps for a ringing in our ears. In contrast, real Christian love never ends. Also, genuine love is the very opposite of selfishness; it blesses all those around us. Notice the description of love as *continuous* in verses 4–8: Love is patient; is not easily angered; always protects, trusts, hopes, perseveres; love never fails. In this world of change, people long for something dependable and unchanging. Our privilege is to show them Christian love and to win them for eternity.

Science Demonstration

Everyone has heard the sound of a gong or cymbal. This activity describes another type of unusual sound device traditionally called a *bullroarer.* We will build a modern copy of this ancient instrument. Bullroarers dating from Old Testament times have been found by archaeologists in England. More familiar musical instruments, the harp and the flute, are mentioned as early as Genesis 4:21 (see lesson 6). Sound and music have always been part of mankind's art and enjoyment.

The instrument can be prepared ahead of time, or participants can make their own. Almost any flat, stiff material will suffice. It initially is cut roughly into a fish shape. If lightweight plastic or cardboard is used, wrap tape around it and add a few paper clips for added weight. A wooden paint stick also works well if it is cut out ahead of time. The size can vary, but the following dimensions give good results: 8 inches long, 3 inches wide at maximum, and about $\frac{1}{6}$ inch thick (see the figure). The object's edges can either be tapered or left flat. A hole is now placed near the wider end. A length of strong string is threaded and knotted through the hole.

Now make sure there is safe clearance and whirl the object in a vertical circle. If the object has sufficient weight, it soon should begin to spin and produce a deep vibrating sound. This may take some practice, and the resulting pitch and loudness depend on the speed of the motion. If the bullroarer refuses to "sing," twist the string 15–20 times before swinging it in a circle. Like a gong or cymbal, the whirring bullroarer sound is impressive and carries a long distance. When the action stops, however, the sound also ceases instantly. How different a brief noise is from the idea of Christian love that continues today and forever.

A centuries-old example of a notched wooden bullroarer found by archaeologists in France. Some aborigines of Australia still use this type of instrument to communicate with each other over long distances.

Science Explanation

All types of sound originate from vibrating objects. Whether a clarinet reed, vocal chords, or a violin string, each can be made to vibrate and thus produce sounds. The bullroarer instrument likewise vibrates. Because it is not perfectly symmetrical, it should spin rapidly when swung in a circle. The sides then beat rapidly against the air, setting up the audible vibration. The resulting deep bass sound has a frequency of about 30 to 40 cycles per second. What we actually hear are pulses of pressure changes in the air that reach our eardrums and cause them in turn to vibrate.

An actual performance of bullroarer playing occurred during a lecture at the Royal Society of London over a century ago, in 1885. Before the era of radio and television, such public lectures were popular entertainment. Here is a listener's description of the early event:

> At first it did nothing particular when it was whirled round, and the audience began to fear that the experiment was like those chemical ones often exhibited at institutes in the country, which contribute at most a disagreeable odor to the education of the populace. But when the bullroarer warmed to its work, it justified its name, producing what may best be described as a mighty rushing noise, as if some supernatural being fluttered and buzzed his wings with a fearful roar. Grown-up people, of course, are satisfied with a very brief experience of the din, but boys have always known the bullroarer in England as one of the most efficient modes of making the hideous and unearthly noises in which it is the privilege of youth to delight.

This is a humorous picture of an earlier time. A century ago, childhood amusements obviously were more creative and less expensive than many today.

24

Lots of Cotton

Theme: God blesses us beyond our imagination.

Bible Verse: *Now to him who is able to do immeasurably more than all we ask or imagine, according to his power that is at work within us* (Eph. 3:20).

Materials Needed:

    Bag of cotton balls

    Small glasses

    Water

Bible Lesson

In our minds it is natural to assume that God's power, though great, is limited. Since we often become weary, we therefore think God must likewise tire. Perhaps this is a hidden reason for assuming that the creation week actually must have covered a long period of time. However, God doesn't need our help in this way. He could have completed his universe in a microsecond, drawing upon his infinite reserves of energy. Instead, however, the creation was accomplished in six literal days, exactly as Genesis states. And God is still able to do more than we can ever ask or think. We should therefore be bold in our prayer requests. This limitless power of God is also within us, through Christ. Ultimate victory for the Christian is a certainty.

Science Demonstration

We will illustrate what appears to be impossible. Show the listeners a small glass nearly filled with water, to within about 1/4 inch of the top. On a table there also should be a large pile of fluffy cotton balls. Ask the audience to guess how many cotton balls can be added to the glass before water spills over the top. It appears that only 2–3 balls should completely fill the glass. However, the actual results are quite surprising. Many cotton balls can be added to the glass, one after another, and no water will spill. Instead, the water is totally absorbed by the cotton, which greatly shrinks in size. Press the cotton balls down firmly together with your fingers to make room for others. I was able to squeeze 100 cotton balls into a small glass in this way. This large capacity for cotton balls is unexpected. Just as surprising are God's abundant blessings that continue daily.

Many cotton balls can be added to an almost full small glass of water without spilling any of the liquid.

Science Explanation

Cotton fibers are largely made of cellulose, the main component of many plants. This complex carbohydrate has the formula $(C_6 H_{10} O_5)_n$ where n may be a large number. The individual cotton fibers are about 1 inch long and only 10–20 microns (10^{-6} meter) in diameter. This is about 10 times smaller than the diameter of a human hair. A fluffy cotton ball of fibers is mostly empty space. Therefore when compressed, its volume is greatly reduced. The water used in the demonstration is for visual effect, and it also helps hold the cotton together in a wadded form. Cotton is the most universal fiber known, used worldwide for making cloth fabric. It is a marvelous gift from the Creator. After the lesson, the cotton balls can be spread out to dry for reuse. However, they will not resume their original fluffiness, even when thoroughly dry.

25

A Bouncing Ball

Theme: All things wear out.

Bible Verse: *They [heavens and earth] will perish, but you remain; they will all wear out like a garment* (Heb. 1:11).

Materials Needed:

Assortment of balls (golf, ping-pong, rubber, softball, Superballs)

Yardstick or tape measure

Bible Lesson

This present world is only temporary. It did not form by itself, nor can it last forever. The earth and the heavens beyond are slowly wearing out. It is an everyday experience that all things eventually need replacing, including bicycles, cars, clothes, and lightbulbs. On a more personal level, each of us grows older. This may be especially noticeable after a long day of work. The degeneration and decay seen in nature are part of the curse. This began when Adam and Eve disobeyed God, as recorded in Genesis 3. The universal trend toward disorder is one of the most basic findings of science. It is sometimes called the second law of thermodynamics (the first law is discussed in lessons 4 and 26). The second law is a direct challenge to the theory of evolution that assumes progress and improvement of living things over time. There have been many attempts to explain how evolution can occur against the universal downward trend, but such attempts are artificial and insufficient.

Our verse states that the only exception to the law of decay is the Person who created all things in the first place. Would you also like to overcome the law of decay and live forever? Then call upon the Lord's name and become one of his children. Eternal life will far outlast this present, temporary world.

Science Demonstration

Participants will observe the bouncing of a ball. Then they will calculate the total distance the ball travels before stopping. The more "bounce" the ball has and the harder the floor, the better. Begin by holding a ball above the floor, perhaps 5 feet high. Then drop the ball and let it continue to bounce until it stops. You will notice that each repeated bounce is somewhat less than the previous height, as expected. This decrease illustrates the tendency of all things to wear out, slow down, and eventually stop moving. No matter how much "bounce" the ball has, its motion will slowly diminish. Participants may drop various balls with similar results: The ball eventually stops bouncing.

One can determine the total distance traveled by a bouncing ball before it stops. This is especially interesting because a ball may make countless tiny bounces in stopping. One needs only to measure the height loss r for the first bounce. If the ball is dropped from 5 feet and rises to 4 feet, then $r = {}^4/_5 = .8$. Golf balls, ping-pong balls, and Superballs generally have an r between .8–.9. Knowing r, then look in the table to find the total ball travel. The table is constructed for 5-foot initial drops. The answers are surprisingly large. The stopping of a bouncing ball is just one illustration of Hebrews 1:11. All things wear out, but the Lord remains forever.

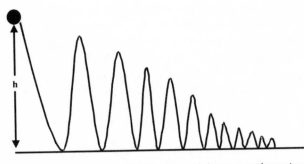

A ball loses some height with each bounce and eventually stops. It is shown here moving to the right but may also bounce vertically.

Science Explanation

In this demonstration the fractional height loss r in each bounce is called the *coefficient of restitution*. If the ball is dropped from a height h, then total travel distance D is given by:

$$D = \frac{h\,(1 + r)}{1 - r}$$

With example values of h = 5 feet and r = .8,

$$D = \frac{5\ \text{ft}\ (1.8)}{.2} = 45\ \text{feet}$$

There are several ways to describe the second law of thermodynamics:

Energy is becoming unavailable.

Entropy (disorder) of the universe is increasing.

Nature tends toward disorder.

The only exceptions to this law involve biblical miracles such as the long-lasting sandals of the Israelites while in the desert (Deut. 29:5). Another is the story of Lazarus, whom Jesus raised from the dead (John 11:1–44).

An object may show temporary increased order at the expense of its surroundings. For example, orderly garden plants may result from sunlight, soil, rain, and the care of a gardener. However, the plants do not form by random chance. Imbedded in their DNA are the instructions for utilizing energy and for growing. Plant growth is not evolution but instead is a result of imbedded design. The spontaneous evolution of life from raw material remains in conflict with the second law.

Height loss (r)	Total ball travel before stopping (feet)
.99	995
.95	195
.9	95
.85	62
.8	45
.75	35
.7	28
.6	20
.5	15
.4	12
.3	9

26

Penny Toss

Theme: Probability rules out evolution.

Bible Verse: *The grass withers and the flowers fall, but the word of the Lord stands forever* (1 Peter 1:24–25).

Materials Needed:

Five pennies for each participant

Plastic cups

Bible Lesson

The two most basic laws in all of science both involve energy (see lessons 4 and 25). The first law says that energy is constant or *conserved.* This means that energy cannot be created or destroyed. It can take many different forms including heat, motion, light, and chemical energy, but the total amount of energy remains constant. This energy law probably was established at the end of the creation week. At that time God ceased placing energy into the physical universe from his infinite supply.

Our Bible verse concerns the second fundamental science law, the tendency of all things to lose their energy to the environment and to wear out. Grass and flowers have a glorious season of growing in the sunlight. Soon, however, their life cycle is complete, and they fade. Death and decay are also expressions of this second law. There is a universal trend in nature for all things to deteriorate and become disordered. This second law probably began or was intensified in Genesis 3 at the time of the fall or curse. In contrast, the Bible stands forever. God's truth remains, while earthly ideas and objects come and go.

Science Demonstration

This exercise demonstrates the tendency of nature toward disorder. Each participant is given five pennies. The coins are then tossed 30 times, keeping track of the number of heads for each toss. It may be easiest to place the coins in a plastic cup, then tip them out on a table. Obtaining either 0 heads or 5 heads for a toss is quite unusual. It happens, but a mixture of heads and tails is more likely. The results should be tabulated. One sample series of 30 tosses gave these results:

Number of heads in a toss	0	1	2	3	4	5
Number of times it happened	1	4	13	9	2	1

Older participants might be encouraged to sketch a *bar graph* of their results. For the numbers shown, the graph looks like this:

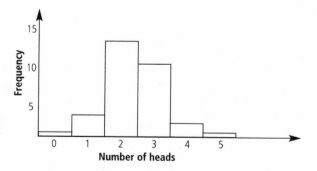

The graph shows that a small or large number of heads is quite unlikely. Most often, one gets 2 or 3 heads. Getting 0 or 5 heads in a toss represents a high degree of order. This is more unusual since nature tends toward the most probable or average condition.

One can compare the coin activity with evolution. This theory assumes that matter somehow came together spontaneously, long ago, to make the stars, planets, and even life itself. However, a simple coin toss shows how difficult it is to produce order. And life is far more complex than five coins that occasionally show all heads or tails. Instead, the chance origin of life would be equivalent to trillions of tossed coins all showing heads simultaneously. It simply will not happen. All life, whether that of plants, animals, or mankind, is a testimony to God's planned creation.

Pennies can be used to illustrate probabilities when tossed.

Science Explanation

In tossing one coin, the chance of getting heads is ½. Here are some probabilities for additional coins:

Coins tossed	Probability of all heads
1	.5
2	.25
3	.125
4	.0625
5	.03125
10	.00098
20	.000001
30	.000000001

In general, for n coins, the chance of all heads is $(0.5)^n$. This number becomes vanishingly small for a large number of coins.

The bar graph for a large number of coins is bell shaped and called a normal or *Gaussian* curve. Our experimental curve of "heads" with just five coins is a rough approximation for such a distribution. A high degree of order, with either few or many heads in coin tossing, falls on the outer edges of the curve and is much more unlikely. And this is the vanishing small realm where the probability of evolution must exist. The amount of disorder of a system, whether coins or atoms, is called *entropy*. Entropy constantly increases as the created universe wears down. The two basic laws described in this lesson are called the first and second laws of thermodynamics. The only exceptions we know of involve miracles. The Creator who established these laws in the first place is able to lay them aside at his will.

27

Sedimentary Layers

Theme: The great flood reshaped the earth.

Bible Verse: *By these waters also the world of that time was deluged and destroyed* (2 Peter 3:6).

Materials Needed:

Funnel

Small container

Sugar (white sugar and a colored sprinkle variety)

Tall clear glass

Bible Lesson

In lesson 7 we explored the geographic evidence for a global flood in Noah's day. Here, Peter tells us that in the last days scoffers will laugh at the flood story. And flood critics are indeed numerous in our day. They suggest that, at most, the Genesis flood story is an exaggeration of a small, localized event. But this is not Peter's view. He explains that the world was completely flooded and destroyed. And the present-day rock layers and fossils worldwide testify to this great event. Likewise, the Lord will someday again cleanse the earth, the next time by fire (2 Peter 3:7). All of us who are on the Lord's side need have no fear of this future judgment. And the invitation of safety remains open to all, by putting our trust in God before it is too late.

Science Demonstration

This experiment concerns the sedimentary layers of rock that cover much of the earth's land surface. They are dramatically visible in places like the Grand Canyon. These layers are considered by many geologists to be monuments to time. The rock layers are assumed to have formed very slowly as sediment collected in ancient seas. Of course, this process was not observed. Instead, however, there is evidence for a rapid formation of these multiple rock layers.

Two materials are needed with similar densities but slightly different sizes. If available, two different sizes of sand grains will work well. Another suggestion is to use two varieties of sugar grains. To distinguish them, one is the colored variety used as a sprinkle for holiday cookies and the other is everyday white sugar. Usually, the colored variety has slightly larger crystals than white sugar. This size difference is important.

Now thoroughly mix the colored and white sugars together in a container, perhaps 1/4 cup of each. Then slowly pour the mixture into a tall clear glass through a small funnel. What should be observed is the partial separation of the sugar varieties into separate layers as the mixture accumulates. Multiple thin layers should result as the two falling materials sort themselves out. If the separation does not occur, try a smaller stream of falling particles.

Since sand grains can similarly separate in a water setting, many of the layers observed in nature may not represent a long time scale after all. Instead, the multiple rock layers could have resulted from the Genesis flood event. In the initial flood stages, much sediment from the earth was stirred up. Then as it gradually settled out, the layers formed quickly. Such multiple layers actually were seen to form during the volcanic eruption of Mount St. Helens in Washington State in 1980. At that time, many distinct layers of volcanic dust and rocks formed very rapidly.

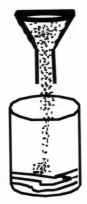

Layering occurs when different-sized particles are filtered into a container.

Science Explanation

Different-sized particles will naturally sort themselves as they settle either in a dry or wet setting. Multiple thin layers or *lamina* have been observed in Mount St. Helens dust deposits (1980), and also in shoreline sediments. The reason for this is not well understood. Somehow, because of size, shape, and density differences, the particles separate. Ongoing research continues to explore this phenomenon. One leading researcher is French scientist Guy Berthault. He has experimented with clay and silt particles in ways similar to our demonstration, both in air and in water. Thin layers are observed to form as the falling particles redistribute themselves. Berthault's work challenges some basic assumptions about the *geologic column*, the major rock layers of the earth. Instead of a monument to time, the rock layers may be a monument to the Genesis flood.

28

Cave Icicles

Theme: Creation was recent.

Bible Verse: *But do not forget this one thing, dear friends: With the Lord a day is like a thousand years, and a thousand years are like a day* (2 Peter 3:8).

Materials Needed:

Baking soda or Epsom salts

Paper towel

Three glasses (2 large, 1 small)

Water

Bible Lesson

Our verse describes God's view of time. He is not controlled by the ticking of a clock as our schedules are. Instead, he is completely above the limitations of time. In fact, God can look down upon history all at once, from the moment of creation to the end of time. Time began at the creation week as a framework for our daily lives.

Sometimes 2 Peter 3:8 is used to imply that a long time period passed during the creation week. In this view, each creation day was actually a thousand years or much longer. However, the verse context is not discussing creation. Instead, it declares that what might seem to us to require a long time can be accomplished quickly by the Lord. The creation *could* have been completed in 6 microseconds or over 6 trillion years. However, from Genesis 1 it is clear that the actual duration was six normal days. After all, the creation account is the origin of our calendar week. And a literal reading of Genesis shows that the creation took place just thousands of years ago, not billions. Alleged evidence for long ages—radioisotope dating, star distances, etc.—all have alternate interpretations. Our science activity will explore just one of these supposed evidences, stalactite formation in caves.

Science Demonstration

Many of us have visited underground caves. Unusual formations found there include stalactites hanging from the ceiling and also stalagmites growing upward from the cave floor. These stone "icicles" grow as minerals are left behind by dripping water from the cave roof. It is often assumed that stalactites require thousands or even millions of years to form. However, this simply is not true. Stalactite growth depends on many factors including cave humidity, temperature, air circulation, the rate of dripping water, and mineral content. When conditions are ideal, cave formations may develop rapidly.

Miniature stalactites can be grown over a period of several days. First, fill two glasses more than half full with either baking soda or Epsom salts. The Epsom salts makes a more realistic stalactite. Then add warm water to the height of the powder in each glass. Stir the solutions to dissolve as much powder as possible.

Twist a paper towel to make a thick ropelike cord about 12 inches long. Place the ends of the towel in the glasses and let the center hang downward as shown, between the glasses. Liquid should creep along the towel from both ends and drip off the center. Place the smaller glass under this drip. Over several days, a small stalactite should form near the center of the towel. When the center glass fills with drip solution, it can be poured back into the large glasses.

The leader may want to grow some stalactites ahead of time to show the group. Materials could then be given to participants to take home.

Miniature stalactites can be grown on a wet paper towel.

Science Explanation

The two alternate chemicals suggested are baking soda, $NaHCO_3$, and Epsom salt, $MgSO_4 \cdot 7H_2O$. Both dissolve readily in water. Capillary action causes the liquid to move along the paper towel, which acts similarly to a wick. Then as the water evaporates, the chemical is left behind to form a growing residue.

Cave formations are typically made of the minerals calcite, $CaCO_3$, or quartz, SiO_2. These natural stalactites are much harder and more durable than our homemade stalactites. Stalactites have sometimes been observed to form very rapidly. They have been found wherever there is seeping mineral water, for example in mines, the basements of buildings, and under bridges. When derived from concrete, the mineral formations often consist of calcium hydroxide, $Ca(OH)_2$, rather than calcite, $CaCO_3$. The hydroxide material is typically soft and flaky. Ongoing creation research demonstrates the formation of the hard, solid $CaCO_3$ form of stalactites as well, under ideal conditions.

Many cave formations probably grew rapidly in the centuries directly following the Genesis flood. For many years there was much mineral-laden water moving through the underground.

The slower growth noticed in most modern caves is probably on a greatly reduced scale from the earlier, rapid growth.

The conclusion is that cave stalactites and stalagmites do not prove that the earth is ancient. Many scientists believe that the earth was created just a few thousand years ago. Many cave formations are not a testimony to time but rather to the flood-waters of Noah's day.

29

Wandering Stars

Theme: Do not follow false teachers.

Bible Verse: *They are wild waves of the sea, foaming up their shame; wandering stars, for whom blackest darkness has been reserved forever* (Jude 13).

Materials Needed:

Masking tape

Pencil

Poster board (dark color)

Salt or sand

Scissors

Small paper cup

String

Two chairs

Yardstick or meter stick

Bible Lesson

The Epistle of Jude warns against false ideas that arose in the early church. One such heresy was called gnosticism. This philosophy saw all matter as basically evil while only the non-material, spiritual realm was declared good. Gnosticism challenges the biblical truth of an original perfect creation. It also falsely teaches that Christ's earthly body was not real, since it then also would have been evil.

Throughout history the stars above have been useful as guides. Stars can reveal one's location on earth by determining the cardinal directions. Stars also provide a useful clock and calendar system for mankind. Stars were especially useful for these purposes before maps, clocks, and calendars were readily available. Certain stars such as Polaris, known also as the North Star or Pole Star, are still especially useful as guide stars.

The words of Jude 13 give a graphic description of false teachers. These leaders who mislead others are called wandering stars. They will quickly lead their followers astray. In like manner, not all objects in the night sky can be trusted to give true direction. Make sure that you follow good leaders who are true to God's Word.

Science Demonstration

We will observe a complex motion that helps illustrate the path of a wandering star. Separate two back-to-back chairs as shown in the figure and secure a yardstick to the tops with masking tape. Cut two 4-foot lengths of string. Attach one string to the stick to form a V-shaped loop hanging downward. Pass the second string through this loop and attach it to the rim of the paper cup with tape. Adjust the setup as needed so the cup hangs about 4 inches (10 centimeters) above the floor. Position a sheet of poster board beneath the cup. Now partially fill the cup with dry sand or salt grains and puncture the cup bottom with a pencil point. Holding your finger over the hole, pull the cup to one side and release it. The falling sand should trace out various complex patterns on the floor, depending on the relative lengths of the strings in the "Y arrangement." Try several different designs, returning the sand to the cup between patterns.

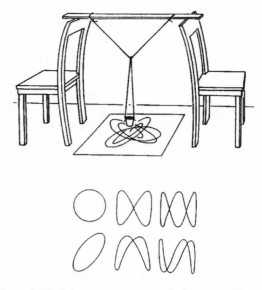

Sand or salt falls from a paper cup attached to a pendulum.
Complex designs result from the pendulum's motion. The lower
drawing shows some resulting patterns.

Science Explanation

The word *planet* comes from the Greek word for *wanderer* as
used in Jude 13. For many years, planets confused observers
with their complex motions across the night skies. These objects
seemed to occasionally slow down, brighten, and even move
backward over a several week period. Johannes Kepler in 1621
showed that the planets follow elliptical paths as they circle the
sun (see lesson 13). Planetary motions are quite predictable once
their paths are understood. Planets are members of the solar
system and therefore are relatively close to us, while the background night stars are very remote. Because of this, the planets
continually change position with respect to the stars. These wandering planets cannot be used as dependable navigational aids.

The path traced out by the sand grains is called a *Lissajous* pattern, named for the French scientist Jules Lissajous (1822–1880).

The interesting designs result when motion occurs in two perpendicular directions, provided in our case by the double or compound pendulum. Lissajous figures are especially useful in electronics, where they are traced out on an oscilloscope screen. Such patterns help in determining unknown frequencies of electrical signals to a very high precision. Like planetary motion, the sand patterns are complex and changeable.

30

The Wonder of Water

Theme: Water displays God's creative handiwork.

Bible Verse: *Worship him who made the heavens, the earth, the sea and the springs of water* (Rev. 14:7).

Materials Needed:

9-volt batteries

Paper clips

Salt

Small clear containers of water

Tape

Bible Lesson

The Book of Genesis tells us that God made everything including the land, the seas, and the heavens above. All things were made supernaturally, from nothing, by God's mighty word:

> For he spoke, and it came to be;
> he commanded, and it stood firm.
>
> Psalm 33:9

Try as we might, we simply cannot comprehend how God created the universe. As Deuteronomy 29:29 explains, "the secret things belong to the LORD." The supernatural is in God's realm alone, being *outside* of nature by definition. For this reason, all attempts to fully explain the origin of the universe, the earth, or the beginning of life itself are futile. In the science world, natural origin theories rise and then fall again, only to be replaced by the latest new theory. Perhaps we should simply praise the Creator for his work and not attempt to understand exactly how he accomplished it.

In studying the present-day creation, we can see some of the building blocks that God used. For example, our verse describes the water that fills the vast seas and also bubbles from the ground as springs. This common chemical called water is not just a simple, formless liquid. Instead, it is made up of two separate elements, hydrogen and oxygen. In the creation of water, God formed these two elements complete with their electron, proton, and neutron components exactly in place. The details of chemistry and physics also were established during the creation week. It is easy to read Revelation 14:7 without giving much thought to the infinite details that were actually put in place. The words from Scripture truly reveal a mighty God.

Science Demonstration

This activity can be done by small groups, or with a larger group using an overhead projector. Water is divided into its component elements, hydrogen and oxygen gases, by using *electrolysis*. This is the name for the electrical separation of the components of any chemical compound, in this case water.

Two straightened paper clips are attached with tape to the terminals of a fresh 9-volt battery. This is the common type of rectangular battery used in devices such as smoke detectors. Avoid touching the paper clip wires together since this "short circuit" may result in a rapid heating of the wires. However, there is no serious shock hazard with such a small battery.

Dissolve a pinch of salt in a small, shallow container of water; the exact amount of salt is not important. This salt will help the water conduct electricity. Now lower the straightened ends of the paper clips into the water. As the chemical reaction begins, many small bubbles of hydrogen gas should appear around the wire attached to the negative button of the battery. The other wire probably will not bubble, but instead will slowly darken as oxygen combines with the metal of the paper clip. By using a clear, shallow container and an overhead projector, the bubbles arising from the water electrolysis can be displayed.

The battery provides the needed energy to divide the water into its component elements, hydrogen and oxygen. Water molecules do not ordinarily break down in nature because the atoms are tightly bonded together. Even in the form of steam, water is still stable. God made water molecules to be a permanent blessing for the earth.

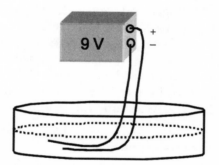

An electric current from a battery will break water down into its component elements, hydrogen and oxygen.

Science Explanation

Many compounds can be divided into smaller components by using electrolysis. The reaction for water is

$$2H_2O \rightarrow 2H_2 + O_2$$

The hydrogen gas is readily seen forming at the negative battery terminal. The oxygen atoms combine with iron on the surface of the paper clip to make iron oxide, FeO:

$$2Fe + O_2 \rightarrow 2FeO$$

If one uses a larger battery or several small ones connected together in series, then enough oxygen may be liberated to cause visible bubbling from the positive battery terminal also.

Donald B. DeYoung is a scientist, author, and professor of physics and astronomy at Grace College, Winona Lake, Indiana. He holds a Ph.D. in physics from Iowa State University and a Master of Divinity from Grace Seminary. Dr. DeYoung is a popular speaker on Bible-science topics for audiences from ages five to ninety-five.